# Our Cosmic Origins

*Paul Dejillas, Ph.D.*

Our Cosmic Origins
First Edition, 2017
Text copyright © 2017 Paul J. Dejillas
Cover design by Cynthia Marquez

Published and printed in USA in 2017 by TATAY JOBO ELIZES, Self-Publisher, under the expressed permission, approval and authorization of the author, PAUL DEJILLAS, who owns the copyright to this book. The author can withdraw or rescind this permission at his discretion without any objection from Tatay Jobo Elizes at any time. Printing of this book is using the present day method of Print-On-Demand (POD) or Book-On-Demand (BOD) systems, where prints will always be available. The copyright owner is free to republish or reprint with other publishers and printers anytime.

ISBN – 13: 978 – 1546319054 &
ISBN – 10: 1546319050

Contact: job_elizes@yahoo.com
Websites: http:tinyurl.com/mj76ccq +
www.jobelizes6.wix.com/mysite

*To all seekers of truth*
*those who have found it*
*those still seeking for it*
*those not seeking for it*

# Contents

# Figures

# Foreword

Yes, the Cosmos is our home. Its magnificence still displays its beauty and grandeur that incessantly bewilder us showering us with its abundance while at the same time depriving us of it, even frightening us with its rage that can instantly obliterate its creation, including us humanity.

What is this world we call our home? What is it in us that deserve both its fury and glory? What is our place in it? The Cosmos we inhabit continues to baffle us as it did to our ancestors millions of years ago.

Did we really originate from the chance fusion of atoms and molecules that produced the chimpanzees, from which species we came from? If so, what makes us different from them? What makes us humans?

I take off from the materialist perspective of quantum physics branching out to other disciplines, cognizant of its relevance to explain our origins beginning from the elements of atoms and molecules that produced other larger-scale structures, including us humans. I realize that quantum physics can serve as the foundation that can connect to all the other sciences like, biology, chemistry, neuroscience, cosmology, psychology, anthropology, philosophy, and theology. Quantum physics can even stand as the most likely candidate to link and integrate isolated studies about our origins in one coherent fashion.

Concerns about our origins have not only come from the province of quantum physics. I make use of the theories and findings of other studies, trying not to deviate from the standards and rigor of quantum physics. ###

# Acknowledgments

This study is a product of more than two decades of research. During this period, I devoured volumes of books in philosophy, religion, theology, economics, anthropology, sociology, parapsychology, neuroscience as well as quantum physics, astronomy, cosmology, biology, chemistry and even delved deeply into the study of the paranormal, the occult, and mysticism myself.

I'm thankful and indebted to many authors; the major ones are named in the References. Their works and ideas greatly contributed to enriching my knowledge about humanity and the Cosmos. I also gained deep insights from highly acclaimed mystics and gurus. I had the opportunity to immerse in the community activities of Buddhists, Hindus, and Muslims – attending rituals in their temples and monasteries, engaging in their meditations, yoga, and liturgical services. I likewise attended several séances performed by mediums and psychics known to be channels of some divine entities. On several occasions, I invited these gifted individuals to lecture in my class and even hold psychic and spiritual sessions with my students.

I had the privilege of interacting with high-profile individuals coming from various disciplines. I am most grateful to Dr. Fred Alan Wolf (a.k.a. Dr. Quantum) for his gracious comments about this study. His insights were most helpful in narrowing down the focus and in clarifying the perspective I am taking in this study. His insights appear here in many appropriate discussions. Without him, this book would not have appeared as what it is now. I wholly accept full responsibility, however, for whatever errors that still lurk in the pages of this study.

I also thank my colleague Emmanuel Ikan-Astillero, Ph.D. who gave his gracious comments on the contents of this book. My gratitude also goes to Psychic and Medium Noel D. Santander, Ph.D. who introduced me into the world of the paranormal. The same gratitude goes to Fr. Benigno S. Beltran, SVD, Ph.D. who motivated me to go into the study on folk religiosity. Warka Adala made me experience the healing energy of the Cosmos while Paul Bouchard gave me the opportunity to personally experience psychic surgery.

Dr. Ava Vivian Gonzales and Sylvia L. Mayuga were kind enough to read previous incarnations of some chapters while Dr. Remedios Nalundasan-Abijan did the final editing of the entire manuscript. My sister Amy Dejillas-Withers spent long hours in facilitating the printing process.

I would like also to acknowledge the contribution and cooperation of my students. It was during my sessions with them that I conceived and presented the ideas and theories discussed here. Their comments were really helpful and this book is a result of an approximately 320 hours of lecture given in a weekly eight-hour session in a span of two semesters.

I would like also to extend my appreciation to the late Atty. Carmen Maluto who encouraged me to do a study that led to the publication of this book.

This work would not have been made possible without the full support and assistance of Dr. Mina Ramirez, President of the Asian Social Institute. Dr. Ramirez, with whom I have had the privilege of association for the past two decades, gave me the opportunity to join the faculty of ASI's doctoral program in Applied Cosmic Anthropology.

Paul J. Dejillas, Ph.D.
For your feedbacks, contact me at:
pauldejillas@gmail.com
Join us in Facebook:
https://www.facebook.com/groups/889112731223672/
Join us in our website:
www.cosmicanthropology.com
March 2017

# Introduction

*Group problems are many; why should there be suffering starvation, and pain? Why should the world as a whole be in the thrall of the direst poverty, of sickness, of discomfort? What is the purpose underlying all that we see around us, and ... what is the key to its present condition?* - Alice Bailey

*... we are the Earth. Everything that gives us shelter and sustenance, all the objects we possess, indeed every atom and molecule of our flesh-bound shells, comes from Earth and will return to Earth. To know our home, then, is to know a part of ourselves.* - Robert M. Hazen

*Standing alone outside my house, I notice clouds merge with other clouds, then break apart before coming together in a cosmic dance. It's still dark but the trees and plants seem to glow with the promise of dawn. I hear and feel the trees sway in the easterly wind, in time to the chirping of birds. I inhale the energy that pulsates around me along with the fresh mountain air and become more alive. I feel somewhat frustrated that I cannot communicate with my surroundings. I want to know what nature feels, what it thinks, what it knows. I can only hope that the vigor and vitality I feel in its presence resonates as an unspoken language it understands.*

*As I float with the clouds, chirp with the birds, sway with the trees, feel the rocks under my feet, I realize that I am not alone. I'm in communion with Mother Nature, throbbing with the power and vitality of the entire universe, the Cosmos. Time seems to freeze and everything is possible. At the mere thought, I can travel anywhere I wish a million times faster than light. All forms disappear, displaced by an awareness of the self that transcends time, space, and matter. In quick succession I merge with the Cosmos, becoming one and the same.*

*I carry the Cosmos within me, yet it carries and engulfs me. The distinction between the Cosmos and me no longer exists.*

*Now there is only eternity, infinity, peace, total awareness. I'm experiencing infinity and eternity. I have become infinity and eternity.*

*I have become peace and awareness. Who's thinking of time when the thinker and the thought have ceased to exist? Who's thinking of time when time has ceased to exist?*

*Who's thinking of space when there is no longer anything to traverse? Who's thinking of forms and matter when there is only awareness of being? The more I reflect, the more I realize that this world we call the Cosmos in which I live is not just a place. It's my home. Yet I am designed for a different realm of life and existence.*

*As I enter the house, a clock on the wall reveals it's six-thirty in the morning. What seemed to be just ten minutes spent outside was in fact more than an hour. Now awakened and energized, I look forward to the rest of my day. ###*

This book is a study of our origin, nature, and evolution. It responds to such questions humanity had been mulling over the millennia. Who or what are we? What makes us uniquely human? Where did we really come from? Did we evolve from the chimpanzees? Or, did some extraterrestrial entities and divine beings create us in their image and likeness? In any case, why did we appear at all? What is our purpose?

What of our future? Are we predestined machines "moved, directed and commanded by exterior influences," as Mark Twain put it in *What Is Man?* or are we free to choose our destiny? Do we totally disappear from the face of this earth and return to dust or is there life after death? If we do live after death, what is it in us that continues to live?

In responding to these questions, I explored the Cosmos. After all, we came out of it; we are made of the same atoms and molecules that produce the stars, star-seed beings, so to say. Our thoughts, feelings, and actions obey the rules that govern the behavior of atoms and the planets. And so is our relationship with our surroundings and with one another. I believe a fuller understanding of ourselves depends so much on our knowledge of this place we live in. But what do we know about the Cosmos?

**Two Competing Views**

When modern science speaks of the Cosmos today, it usually

means the physical structure of the vast, large-scale universe: the macro-cosmos (the clouds, stars, planets, suns, and moons in our solar system, the galaxies, along with dark matter, dark energy, wormholes, and black holes); the micro-cosmos (the world of atoms and their sub-particles); and those in the meso-cosmos (the atmosphere and biosphere) in between.

Mainstream thinking is bent on studying the origin of the Cosmos— its nature and the various forces or laws that govern its behavior. It investigates how the strings, quarks, and leptons evolved into atoms and molecules, which after billions of years, gave birth to the clouds, stars, suns, planets, as well as the plants, animals, primates, and the human species. But viewing the Cosmos this way is most limiting and incomplete. Underlying this view is the deep-seated philosophy declaring that the Cosmos and everything in it is purely material, composed of solid particles, governed by physical forces.

I'm not saying that this view of the Cosmos is in error or false, but our collective knowledge of the Cosmos also tells an entirely different story. For millions of years, our ancestors have forewarned us that there's something beyond what our physical senses can detect. They knew that at its most fundamental level, there's something even subtler than the realm of atoms and subatomic particles. The Cosmos we inhabit contains other realms

## Why this Study is Significant

*Various authors have addressed the significance of viewing ourselves from the cosmic perspective. Neil deGrasse Tyson, astrophysicist and director of New York's Hayden Planetarium declared the importance of continually exploring our Cosmos (2007):*

*During our brief stay on Planet Earth, we owe ourselves and our descendants the opportunity to explore—in part because it's fun to do. But there's a far nobler reason. The day our knowledge of the cosmos ceases to expand, we risk regressing to the childish view that the Universe figuratively and literally revolves around us.*

*In that bleak world, arms-bearing, resource-hungry people and nations would be prone to act on their "low contracted prejudices." And that would be the last gasp of human enlightenment — until the rise of a visionary new culture that could once again embrace the cosmic perspective.*

*Physicist Gary Moring (2002) supported this view. He advanced the idea that the more we know about the Cosmos "the more we will know about ourselves and how we may fit into the grand scheme of the cosmos." Similarly, Joel Primack, Professor of Physics, and his wife Nancy Ellen Abrams, Cosmology and Culture teacher at the University of California-Sta Cruz (2006:11) expressed the significance of understanding humanity from the cosmic perspective:*

*Thinking cosmically might help us experience what seems to be the human part of the universe. Though people tend to focus on their differences — classifying others into us and them — when we humans confront the universe all differences among us become trivial: we vary no more than pearls on a string compared to what's out there.*

*And we pearls may be far more cosmically rare and precious than most people realize. It's only because humans are all bunched together on the planet that we fail to see how extraordinary we are.*

*In a time and world where humanity is too immersed in finding solutions to unprecedented global upheavals, finding our meaning from the perspective of the Cosmos can still offer new outlooks and insights that can guide us in transforming our world. Tyson zeroed in on this transformative effect (2007):*

*The cosmic perspective enables us to see beyond our circumstances, allowing us to transcend the primal search for food, shelter, and sex. The cosmic perspective reminds us that in space, there is no air, a flag will not wave—an indication that perhaps flag-waving and space exploration do not mix.*

*The cosmic perspective not only embraces our genetic kinship with all life on Earth but also values our chemical kinship with any yet-to-be discovered life in the universe, as well as our atomic kinship with the universe itself.*

## My Approach to the Study

I take note of the various studies done by scholars from different disciplines. Their studies, though separate, identify several cosmic dimensions that have their bearing on us. I noticed four major realms discussed in these studies, namely, physical, mental, psychic, and Consciousness realms. I take off on the side of quantum physics cognizant of its relevance to explain our origins. But concerns about our origins have not only come from the province of quantum physics. Other disciplines have their take on our origins too. This is the reason why I explored what the other disciplines say about the Cosmos and weaved them into the discussion, without disregarding the standards and methods of science. This makes this study cosmic in its perspective and interdisciplinary in its approach.

Three daunting methodological concerns confront this exploration, however. First is the highly elusive and puzzling nature of reality.

Oftentimes I have thought I had captured the reality I'm investigating, but digging deeper into its subtleties only makes it more elusive. I find myself soaring up and down into the abyss of darkness, nothingness, and the unknown. Exploration then has become a challenging voyage that led me, at times inspired, to chase the speeding reality, appearing now and then like an ungraspable rainbow, reminiscent of Democritus's experience 2,500 years ago. It is as though I were holding a chunk of ice in my hand one moment seeing it melt slipping through my fingers and vanishing into thin air the instant I feel the ice.

Thus is the changing nature of reality. After all, the gas in the atmosphere turns back into water after being nurtured inside the wombs of clouds before it falls as rain, only to be frozen back into a chunk of ice I can hold in my hand. This is repeated throughout the Cosmos as the explorer's endless chase of an ever-changing reality also seen in the periodic, if slower, changing of the seasons—summer, fall, winter, and spring, wet and dry—or, in a cycle of birth, maturity, decay, and death found in all of life.

In this exploration, the critical reader may wonder: What, then, is the essence of the Cosmos so elusive and seemingly inapprehensible? What is its real ultimate nature?

The second thorny issue confronting this study is the breadth and depth of the totality of the Cosmos. Getting a clear and coherent picture of its entirety can be utterly intimidating and at times frustrating, somewhat like the epic journey of Gilgamesh in search of his divinity and immortality. Even as I think I possess a rational, intelligible view of its nature and behavior, in the end I realize that what I have is almost always incomplete, confusing, jumbled.

Today, the more scientists delve into the question of what the universe is made of, the more they're dragged into the abyss of nothingness, darkness, and mystery. The matter and energy we know today constitutes only five percent of the total mass of the Cosmos. The greater bulk—26 percent dark matter and 69 percent dark energy—remains a mystery (Kathrine Freese, 2014; Scott M. Tyson, 2011). But these elements work in opposite directions.

Like gravity, dark matter tends to attract and pull large objects together but its force is much weaker than the gravitational pull. On the other hand, the force of dark energy acts in the opposite direction. Instead of holding all celestial bodies together in their proper orbits, it pushes them apart (Elizabeth Quill. ed., 2016). Its presence was detected when it was observed that the expansion of the universe is accelerating at a much faster rate than what gravity and dark matter would have allowed.

The challenges are staggering, indeed! For why start the exploration at all, when we can only learn so much about the Cosmos? Still, unlike Gilgamesh who failed to redeem his divinity and immortality, Sisyphus encourages us. Committing himself to push the rock up the hill, he was in the end rewarded at least for his effort.

Today, modern science tells us not to let this formidable task dismay us. It has discovered something it believes can get us out of the cul de sac—the hologram. A hologram is an image in which each part of the system contains information about the whole. Albert Einstein's protégé, Michael Talbot, remarked that even a grain of sand contains information about the entire Cosmos.

The hologram's discoverer, David Bohm, declared: "Any part of the hologram contains information about the whole hologram; any part of the universe contains information about the totality of the universe. There are no differences between the part and the whole." Quantum Physicist Erwin Schrödinger, a contemporary of Bohm, echoed this theory: "I am a piece of existence, but I am also whole." The discovery of the hologram makes the study of the Cosmos not as daunting as it seems at first, with its proverbial light at the end of the tunnel.

This is not the end of our hurdle. There's a third concern and this is the issue of linkage and integration. How do we treat and handle all this information, considering that these cosmic realms have their own distinct worlds obeying rules that may not necessarily apply to the other realms? How can we link particularly the physical and metaphysical realms? Mainstream thinking today is that the two are incompatible; there is no way they can be linked.

This is exemplified in the opposing beliefs that Science and Religion hold on the nature and behavior of the Cosmos. Science contends that the Cosmos is physical, governed by the forces of nature. Modern science says we derived our knowledge about reality based on the study of the physical world. Hypotheses and theories are verified through empirical observation, experimental testing, replications of laboratory experiments in various settings, utilizing the consistency of logic and the precision of mathematics.

Open to peer reviews and critical attacks, modern science is willing to abandon its theories if these do not conform to what is happening in the observable world. It will discard its ideas in the light of new empirical discoveries. In the view of Science, theories—and truths for that matter—are relative, depending on whether they fit the reality outside or not. Even then, only those found most useful and adequate are kept. What ultimately stands out as the most definitive and valid explanation becomes universally accepted by the whole scientific community, without discounting the eventuality that this could still be challenged and replaced as time and conditions change.

Religion, on the other hand, believes in a metaphysical world governed by non-physical laws. Unlike Science, it puts its trust on faith, revelation, in truths known through its sacred scriptures, and on the pronouncements of their respective priests, ministers, prophets, gurus, sages, rabbis, imams, or masters. Those who do not subscribe to proclaimed truths are "in error" (Kurtz, 2003).

For more than four centuries now, the methodologies of Science and Religion have remained incompatible, making integration of the two near impossible. Can a new understanding of the Cosmos and humanity then emerge by combining these two divergent views?

To me, the possibility is great. Today, we witness an emerging trend between and among disciplines towards building links and integration. The academic disciplines are broadly categorized into three branches—physical sciences, social sciences, arts and humanities (see Fig. 1.1).

All study humanity, but differ in their fields of specialization. The physical sciences focus on humanity's origin, structure, evolution, and physiology, while the social sciences focus on the distinctive forms of human relationship. In the meantime, the arts and humanities focus on the great achievements of humanity. A similar categorization can be said of the study of the Cosmos.

In the past few decades, a growing number of social scientists have left their narrow confines by joining their kindred—sociologists, economists, political scientists, anthropologists, and psychologists—and their close relatives—philosophers, humanists, artists, historians, theologians, and occultists—as well as their distant cousins—the physical scientists, physicists, astronomers, cosmologists, biologists, chemists, and neuroscientists.

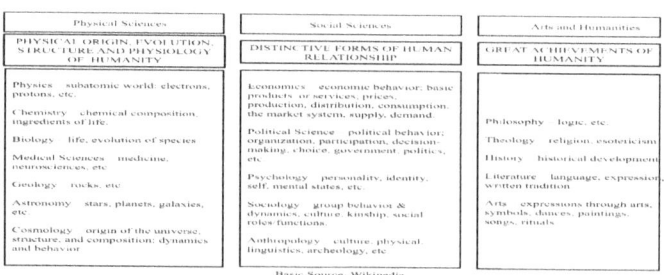

Figure 1.1. Physical Sciences, Social Sciences, Arts and Humanities

Economics, for instance, belongs to the "behavioral sciences" or "social sciences." Much has already been done by economists in branching out to their closest kin like Adam Smith, Thomas Malthus, David Ricardo, John Stuart Mill, and John Locke, who were also political scientists, social scientists, and philosophers. Since the social sciences consider the study of human nature and behavior their province, they see no serious difficulties in bridging their disciplines. The economist-philosopher John Stuart Mill (1859) declared the value of interdisciplinary studies, no less. In his words:

> *The only way in which a human being can make some approach to knowing the whole of a subject is by hearing what can be said about it by persons of every variety of opinion and studying all modes in which it can be looked at by every character of mind. No wise man ever acquired his wisdom in any mode but this.*

Moreover, for a long time now, interdisciplinary studies have been the hallmark of anthropology, in fact considered by many as the forerunner of utilizing multidisciplinary approaches (see Fig. 1.2). Anthropology's encompassing breadth is also found in the arts and humanities studying religious beliefs (Anthropology of Religion) and civilizations unearthed in cave paintings and fossil remains (Archeology). It's also into Ecology (Paleoanthropology and Palynology). Through the years, anthropology has done great service in other fields of human endeavor. It has pursued endless excavations discovering artifacts based on leads coming from ancient myths and legends.

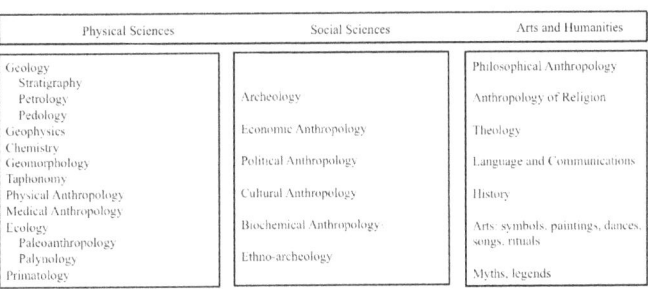

Basic Source: Harry Nelson and Robert Jurmain (1982)

Figure 1.2. Anthropology and Its Branches

Among the many academic disciplines, it is the most interdisciplinary and holistic in its approach. Many used to say that traditional anthropology has no meaning and relevance at all in responding to our day-to-day problems. This is no longer true (H.G. Barnett, 1972:484; Lisa R. Peattie, 1972:487-493; John van Willigen, 1986; and Carole L. Crumley, 2001).

We now witness its active role in the courts of law and halls of justice (forensic anthropology) as well as in the field of medicine (paleopathology, skeletal biology, or osteology). Its more recent addition to the family is "bio-cultural anthropology," a contraction of "biology" and "culture." This new branch investigates both our biological traits (race, color, genes, blood, cranial capacity, brain, etc.) and cultural behavior at various historical time periods and places.

On the same interdisciplinary wavelength, many emerging subjects in the social sciences have appeared today. They are offered over the Internet, advertised in the curricular offerings of schools or integrated into their syllabi and instruction materials. Examples are socio-biology or bio-sociology, neuroeconomics, eco-psychology, behavioral economics, mystic economics, spiritual politics, and hypnotherapy, among other nuances.

It's also interesting to note that, starting in the 20th century, the physical sciences, particularly cosmology, astronomy, physics, biology, chemistry, neurobiology, and the medical sciences have been undertaking joint interdisciplinary studies, crossing beyond the horizon of their respective disciplines in search of new insights. Let me cite the case of Fritjof Capra, a physicist involved in theoretical research in particle physics. Capra (2000) observed similarities and parallelisms in the theories of modern physics and the beliefs of the world's oldest Eastern religions.

Another case is the Anthropic Principle that came out as a result of the studies done by a mixed breed of scientists and humanists. It theorizes that there must be some kind of intelligent force responsible (Brandon Carter, 1974).

Then, there's the Intelligent Design theory advanced by Biochemist Michael J. Behe and his associates. In the humanities, Rev. Pierre Teilhard T. de Chardin's evolutionary view of the Cosmos and humanity linked scientific theories, philosophical speculations, metaphysics, theology, and religious teachings in his concept of the Noosphere.

In August 1987, the Santa Fe Institute, a think tank founded in the mid-1980s, held a conference attended by an eclectic gathering of eminent physicists, economists, computer scientist and biologist. Other such interdisciplinary forums could be ongoing at this writing, their outputs probably waiting for publication. Another institution engaged in interdisciplinary dialogue is the International Conferences on the Unity of the Sciences (SUC) which holds annual gatherings of experts from various disciplines "from the natural sciences to the social sciences and to the humanities… in art, philosophy and religion" (Mitchell M. Waldrop, 1993).

In the final analysis, the knowledge we may derive from our exploration may become even more exciting due to the cosmic approach and method we are utilizing. There are, of course, trade-offs and compromises that may have to be tolerated. Specializing only in one discipline, as the reductionist approach does, always faces the risk of missing what the other sciences and fields of disciplines may have to say about a certain subject.

At the same time, a multidisciplinary approach, while offering a much broader perspective on the subject, may also run the risk of being too general and may lack lots of details which the reductionist offers. Will Durant (2011) may have rightly observed: "…the probability of error increases with the scope of the undertaking, and any man who sells his soul to synthesis will be a tragic target for a myriad merry darts of specialist critique."

On the other hand, too much reductionism without any regard to holism, universalism, and cosmism only isolates us more and more from each other, while too much holism and interdisciplinary approach will continue to be highly vulnerable to being labeled as too abstract, vague, and lacking the fundamental basis that can only be unearthed deep in the ground by the reductionist approach.

Ultimately, what this book lacks is largely dependent on whether the reader leans more towards the reductionist or interdisciplinary approach. But over the past decades, more and more findings have been brought out birthed from both excavated records of the ancient past and new discoveries emerging from modern science. They can give us enough material to sort things out and make sense of the reality we are in today.

I am biased in favor of the holistic and interdisciplinary approach and for good reasons. Michio Kaku (1997) noted that reductionism is waning and has become outdated since it complacently ignores developments in the other fields of endeavor. In its place, we are witnessing an ever-quickening pace in the use of the new method of synergy.

## From Darkness to Light

In retrospect, my exploration was an adventure into darkness and the unknown, uncertain of what I would discover. I sifted through voluminous materials that, at first glance, seemed chaotic, disorderly, and confusing. They ranged from the scientific to the mythical, philosophical, and theological, archaeological, and anthropological. I was immediately reminded of the chaos and disorder that marked the early inception of the universe.

Humanity's attraction to the Cosmos is dated as far back during the time of the Hominids around four million years ago. Archaeological findings of fossil remains, cave paintings, megaliths, monumental edifices, mysterious landscapes, art, and hieroglyphic clay tablets of this period give us a glimpse of how our ancestors viewed the Cosmos and humanity. The earliest available writings are from historical records found in the ancient creation myths of Sumer, Babylon, Egypt, Greece, India, and China.

Other sources include those compiled in the various writings of the world's ancient religions and their sacred scriptures—Judaism, Hinduism, Buddhism, Confucianism, and Taoism, among other pre-Christian religious groups. I also highlight legendary tales on the creation of the Cosmos and humanity like the Sumerian's Enuma elish, and the legendary journey of Gilgamesh. The importance and value of myth in our understanding of the Cosmos and ourselves is underlined by Joseph Campbell (in Keiron Le Grice. 2011):

> *It would not be too much to say that myth is the secret opening through which the inexhaustible energies of the cosmos pour into*

*human cultural manifestation. Religions, philosophies, arts, the social forms of primitive and historic man, prime discoveries in science and technology, the very dreams that blister sleep, boil up from the basic, magic ring of myth.*

Keiron Le Grice also advanced that mythical recordings of humanity's history formed our basis for understanding the present and the past and "have even given rise to the world's major religions, which have in turn shaped and sustained the world's great civilizations." Myths have served the role of explaining our relationship to the Cosmos, to Nature, and our fellow terrestrial beings. They have strengthened our preservation and bond with each other. Suffice it to say, the accounts of ancient civilizations, indigenous peoples, and ancient religions are as much an essential part of our story as the theories and beliefs of modern science.

The field of philosophy, likewise, deserves space in our exploration. Much of our philosophical knowledge relied on the Greek philosophers. The discovery of the atom by Leucippus and Democritus gives modern physics its starting point. Their concept of the nature and behavior of a geocentric cosmos persisted for more than 2000 years into the Medieval Ages.

In this period, I look into the philosophical works of Francis Mercury van Helmont and Gottfried Wilhelm Leibniz, which reveal the nature and dynamics of our Cosmos based on logical reasoning. They differ from the new breed of modern philosophers, known as philosophers of science, who have emerged in our times. I also explore the works of classical scientists, Albert Einstein as well as the founders of quantum physics.

Unavoidable questions about the explorer's objectivity will be raised. In such a process, the researcher must decide which material is relevant and which is to be disregarded; subjectivity and biases inevitably percolate, whether consciously or unconsciously, during the entire voyage of exploration; but this caveat, likewise, applies to anybody who endeavors to study any reality.

The history of cosmology is continuously churning out theories that conflict each other, only to be abandoned at a later time. Such a process will go on unceasingly until a more coherent portrait of the Cosmos is drawn. This spirit prompted this book to discover such a portrait.

**Organization of the Study**

Physicists have always been concerned with finding the ultimate reality that could describe the Cosmos in its totality. Their efforts have led to the birth of the various field theories. These include the following: (1) Materialist Field Theory; (2) Classical Field Theory; (3) Gravitational Field Theory; (4) Quantum Field Theory; (5) String Field Theory; M Theory; and (6) Zero-Point Field Theory. These theories led me to formulate the Cosmic Energy Field (CEF) Theory as a continuation to the search of the grand unified field theory. I discuss these theories in Chapter 1.

In Chapter 2, titled "And the Cosmos Dances to the Tune Played by the Pied Piper," I explore the quantum world to know its characteristics and behavior. After the Big Bang 13.7 billion years ago, these tiny particles coalesce and unite together to form more complex structures.

These structures in turn led to the birth of the clouds, stars, planets, galaxies that eventually produced the necessary ingredients for our solar system to appear. Moving forward in time, I learn how our solar system produced the necessary conditions and elements like air, fire, earth, water, to prepare for the appearance of life and our coming.

In Chapter 3, I focus on the physical realm of creation. I discuss several creation stories from the materialist perspective of quantum physics, trying to weave all of them in a coherent manner to form a bigger picture of the beginnings and development of the Cosmos. I conceptualize creation into seven moments paralleling the biblical account where God created the universe in seven days. With the appearance of life, the cosmic stage is set for our coming.

In Chapter 4, titled "Let us Make Man In Our Image and To Our Likeness," I discuss how humanity appeared, tracing our lineage from the time of Adam and Eve down to the 12 tribes of Israel. Again, I came across several versions of the story. I explored how the Darwinian theory of evolution explains our birth. I also surveyed what the ancient accounts tell us of our appearance.

In Chapter 5, with the title "The Story of Our Civilization," I narrated how we were transformed from scavengers and farmers into scientists, theologians, philosophers, and economists producing the likes of Isaac Newton, Albert Einstein, Nicolas Tesla, Max Planck, Mozart, Beethoven, Rene Descartes, and Immanuel Kant and how we were able to build megalithic structures, computers, space ships that enable us to send men and women to the moon and outside our Planet Earth.

In Chapter 6, I continued my journey into the mental realm. I theorize that the achievements we have accomplished today and in the future lies in our fully developed mental faculties, our brain and mind. It is this capacity and power of our brain and mind that distinguishes us from the chimpanzees we once were. It is in this mental realm that brought our civilization to what we are today—the age of science and technology, industrial revolution, and globalization.

I examine the structure of our brain and how it operates. I also investigate the human mind, its structure and operations. Realizing that they all behave according to the rules of physics, I examine the concept of free will in some detail.

In Chapter 7, I continue to explore our inner world. I discovered that another mysterious world exists out there, which forms part of the Cosmos. It is mysterious because it defies the rules of physics. This world I am referring to is the psychic realm, the realm where paranormal events happen. From my research, I realize that events occurring in the psi realm influence our search for meaning and purpose in life. We cannot ignore this realm otherwise we will be missing one of the opportunities that could make us fully and truly human. I discuss this realm from the perspective of classical science, quantum physics, as well as religion and mysticism.

But there are still other dimensions higher than the physical, mental, and psychic realms. I am referring to these higher dimensions as the Consciousness realm, which I discuss in Chapter 8. Like the psychic realm, three views outshine, namely, the classical, quantum, and metaphysical paradigms. But out of the synergy of these paradigms emerges a new paradigm, which I baptize as the New Consciousness Paradigm.

As an epilogue, I present the process and effect of my journey under the title "In the Spring of Life: The Dawning of an Awakened Humanity." I highlight my finding that the cosmic evolutionary process has so far led us to the dawning of an awakened humanity. With humanity's newly found personality, everything in the Cosmos springs to life again, now, with an even greater vigor and vitality, to face another uncertainty that the future may bring on us.

I invite the readers to explore this place, still unknown and undiscovered by humanity. I do not promise anything but the prospect of amassing new insights about ourselves as well as our surroundings, societies, and the world in general. Viewing ourselves from the cosmic perspective might still give us fresh ideas and programs that could more meaningfully and effectively respond to the harsh conditions we are in today.

# 1

# In My Father's House Are Many Mansions

*In my Father's house are many mansions: if it were not so, I would have told you. I go to prepare a place for you.* - John 14:2

*In our universe we are tuned into the frequency that corresponds to physical reality. But there are an infinite number of parallel realities coexisting with us in the same room, although we cannot tune into them.* - Steven Weinberg

*Up to the present, the various kinds of things existing in nature have ... been found to be organized into Levels. Each level enters into the substructure of the higher levels, while, vice versa, its characteristics depend on the general conditions in a background in other levels both higher and lower, and in part in the same level.* - David Bohm

*"The universe is a 'system' composed of many parts or components. System parts are described and discussed independently, but they are all interrelated and operate as one unit. It takes all the parts to make up the universe; all parts are essential and absolutely necessary for the universe to work as a unit. If any one part were to be removed, the universe would not work. "Because all the components of the universe are so interrelated, it is difficult to write about any one component without continually making reference to other components ...* - Florencio J. Perez

*There is something 'out there,' some energy field, that allows our thoughts to propagate through space—even over vast distances. The implications of this are stunning—particularly when you start to consider how it seems that every living thing in nature is listening to everything else.* - David Wilcock

*At our most elemental, we are not a chemical reaction, but an energetic charge. Human beings and all living things are a coalescence of energy in a field of energy connected to every other thing in the world. This pulsating energy field is the central engine of our being and our consciousness, the alpha and the omega of our existence.* - Lynne McTaggart

*I usually spend my time unwinding with my circle of friends— unfortunately or fortunately all foreigners, expats with varying fields of specialization and careers—at our favorite watering hole, enjoying wine with cheese and bread before sharing a sumptuous dinner.*

*These get-togethers often end just before midnight, which makes it easier for me to return to my home in the mountains, when city traffic disappears. I tell my friends that despite my hectic schedule or teaching load, I nevertheless have a refreshing weekend that makes me look forward to seeing my students again. It was only much later I realized that cosmology (the subject I focus on in my anthropology class) and wine have become inseparable. On certain occasions, bottles of wine, tarragon teas, and assorted pastries brought by my students invade my classroom during short breaks.*

*But sometimes my weekly "unwinding" becomes a "rewinding" of what transpired in class. Whenever this happens, the gates of hell break loose as my fellow barflies try to extinguish my remaining resolve.*

*Take what happened one evening, when what started as a jovial conversation suddenly turned sour. I felt I was on the witness stand surrounded by a panel of jurors, a judge, and a crowd of spectators in a hall of justice. It was my favorite watering hole, but it was as if I had my right hand raised in semi-surrender and taken an oath to "tell the truth and nothing but the truth, so help me God." I was probed by a prosecutor, with no lawyers to defend me.*

*All I had was the hope that my invisible guides and guardian angels would come to the rescue. I later found out that, in the heat of the interrogation they did so without my being aware of it.*

**FRIEND (F):** *What do you teach?*

***MY RESPONSE (P):*** *Anthropology.*

*F: What particular subject in Anthropology?*

*P: The study of the Cosmos.*

*F: This is the first time I have ever heard about?*

*P: It's really a new way of looking at humanity from the perspective of the Universe and trying to make sense of opposing findings by several disciplines, coming up with a story or model that accommodates most, if not all, of the findings so far, including those yet to be discovered.*

*F: Hey, that's really something! Let me get this right—in your class human beings are studied from the perspective of the universe, stars, and galaxies.*

*But what do the sun, moon, stars, planets, and galaxies have to do with us, more so our brains and minds?*

*P: The mental realm is part and parcel of the structure of the universe. Hard sciences like biology, chemistry, and neuroscience are now immersed in the study of the mind, thoughts, emotions, free will, and consciousness—which used to be the exclusive domain of the soft sciences.*

*The study of the mental world has now become largely interdisciplinary; everybody's talking about it. It's now a thriving industry with studies and research funded by giant institutions and good-intentioned philanthropists who have their own stakes in the matter.*

*F: What do you mean exactly by the mental world being part and parcel of our Cosmos? What does the mental world do to affect our lives?*

*P: People across disciplines are slowly realizing that a connection exists between the physical and mental worlds. What you think—your inner thoughts—create objective, external reality.*

*So many things we enjoy now like electricity, airplanes, television, cell phones, iPods, Kindles, the Internet, Wi-Fi, Google, Facebook, and other products of modern technology were first incubated in the minds of their discoverers and inventors.*

*F: Yes, now I see what you mean, but I have difficulty with the claim that the stars influence our life and behavior—if you believe they do, does this mean you're also into Astrology, in star and Zodiac signs? How about mental telepathy, ESP, dream interpretations, clairvoyance, communications with the dead, bending spoons, and even communication with spirits and the dead?*

**P:** *Well, for over a few decades now, anthropology has been into all the sciences. I don't see any reason how the inclusion of the paranormal world into the study of the Cosmos is anything new at all.*

**F:** *It just seems too ambitious for anthropology to include it. Isn't the paranormal the domain of psychology and psychiatry?*

**P:** *Each discipline is all trying to explain psychic abilities and events. Quantum mechanics has given us plausible explanations for paranormal phenomena. Quantum physicists are now talking about mind reading, clairvoyance, invisibility, immortality, bi-location, astral projection, near-death experiences, and replicating them in their labs.*

**F:** *Don't tell me you're also into occultism, spiritualism, and mysticism!*

**P:** *As a matter of fact, I invite psychics, mediums, healers, mystics, and spiritualists to speak in my class. They request my students to bring spoons or objects memorable to them. These guests arrive with crystal pendulums, paintings or poems inspired by spirits, and bottles of energized water. My students and I also make pilgrimages to Buddhist temples and Hindu ashrams. We visit the "clinics" of cosmic energy healers, pranic healers, psychic surgeons, and hypnotherapists.*

**F:** *That doesn't sound like any subject I had in college; your students must be very open. Let me see, so far you've mentioned the mental, physical, and psychic worlds. Is there anything else I missed?*

**P:** *There's one more—the world of Consciousness. It includes self-awareness, interconnectivity, interrelatedness, and oneness among cosmic residents. Consciousness is union with the Pure, the Absolute, the Ultimate Principle. It's the realm of the Spirits -- God, also referred to as the "Cosmic Mind and Intelligence," "Intelligent Designer," the "Matrix of All Reality."*

**F:** *In other words, Consciousness is the field of mystics, saints, and the religious, but certainly not of scientists like Albert Einstein, Carl Sagan, and Stephen Hawking!*

**P:** *On the contrary, those scientists you just mentioned recognize Consciousness. Albert Einstein and Carl Sagan acknowledged a God of elegance, beauty, and order, while Stephen Hawking speaks of a* *non-interventionist God, who set everything in order from the very beginning and therefore no longer necessary for the ordering and sustenance of the Cosmos. There are scientists deeply into Buddhism, Hinduism, Judaism, Kabbalah, Christianity, Islam, and other world religions, including Theosophy, Occultism, Maharishi, and the rest of the so-called New-Age movements. Their ideas about God can be enlightening and instructive as well as perplexing and mystifying to ordinary people like us.*

*F: So what does quantum physics say about a world inhabited by God?*

*P: The answer so far isn't to be found along religious lines. Physicists are talking about parallel universes and multiverses instead of a universe. Their parallel universes are as strange and mind-boggling to us as the paranormal and mystical worlds are to them.*

*These parallel worlds are believed to be inhabited by extraterrestrial beings from civilizations much more advanced than ours.*

*F: You mean ETs, UFOs, alien abductions, ancient astronaut theories?*

*Aren't they the favorite topics of Steven Spielberg, the National Geographic, Discovery and History Channels? All in all quite entertaining, perhaps even instructive, and definitely a guaranteed blockbuster at the cinemas, but surely nothing more than that!*

*P: A recent theory in quantum physics—the so-termed M-theory that sprung from the String Theory — talks of parallel universes or the existence of other dimensions beyond our four-dimensional space-time. These extraterrestrial universes are in the other dimensions of reality. The current theory has at least 11 dimensions, but many scientists hold that there could be an infinite number of them, which remain to be discovered through mathematics.*

*M-theory suggests that these extra dimensions of reality could in fact be populated by advanced beings possessing higher intelligences, with technologies that can travel faster than the speed of light. The UFOs sighted again and again over the past decades everywhere around the world could be the inhabitants of these extra dimensions of life.*

*F: Seriously? Are you talking about an impending invasion of Planet Earth?*

*P: It is serious—the possibility of extraterrestrial life, even invasion, is what compelled scientists at the National Aeronautics and Science Administration to send messages into space, build time machines, space stations, and launch space ships to Mars.*

*NASA scientists are now perfecting the art and science of teleportation, bi-location, and remote viewing to explore and spy on other advanced civilizations. Billions of dollars in research grants have been spent on them.*

*And efforts are now focused on creating technologies to inhabit other planets in our galaxies.*

*F: But how do they expect to go into other universes?*

*P: Scientists talk about black holes, wormholes, dark energy, and dark matter. They theorize that these are galactic portals UFOs use to enter and exit from one dimension to the next, including our Milky Way galaxy.*

*F: Like I said, psychic powers and aliens are guaranteed box-office hits, but in the end I think most people are just content with their own little worlds, their basic way of doing things, which, at least on this planet, continues to be about survival, prestige, and power. They're probably wondering what they have to do with the Cosmos!*

*P: In my classes, I propose a theoretical model of one Cosmos consisting of at least four interconnected dimensions. As in a properly functioning system, each dimension is a piece that can neither operate nor exist outside the system, no matter how isolated and insignificant it may seem. In turn, all the bits and pieces in these dimensions are essentially connected to the integral system. Everything is innately related and connected to each other since everything draws life and energy from the one Big Source we know as the Cosmos.*

*F: So I guess the answer to what folks concerned with survival have to do with the Cosmos is "plenty" — whether they're aware of it or not. Well my friend, let's drink to that!*

*Meanwhile in the center stage, the TV starts feeding the live football game between Germany and Spain. Everybody in the bar falls quiet, intently listening to the commentators' blow-by-blow account. ###*

There had been several attempts to explain the Cosmos in its totality. Until now, no such attempts have yet been attained to qualify as the "Theory of Everything" (TOE). Science, for one, focuses its concern mainly on the physical and material aspects of the Cosmos, while Religion and Mysticism give so much attention on the metaphysical and spiritual realm that hints of the existence of an external Creator, considered as God, gods, or goddesses.

Nevertheless, there had been successful efforts in the past to introduce a theory that could connect the physical and the metaphysical realms. One such attempt was the work of French Teilhard de Chardin (1881-1955). Teilhard, as he was more popularly known, could be very much advanced and ahead of his peers. He was a bit reclusive preferring to be alone using most of his time in search of a theory that could explain the what, how, and why of everything.

He was both a theologian and a devoted Jesuit priest earning his baccalaureate degrees in philosophy and mathematics. But he was also trained as a paleontologist and geologist while dabbling as a professor in physics and chemistry at a Jesuit College in Cairo, Egypt.

Teilhard was an example of an individual going beyond the confines of his disciplines and entering the realm of the physical sciences where his fame rose because of his participation in the discovery of the Peking Man.

It was his exposure in the physical sciences that he was able (or at least made an attempt) to bridge Science and Religion by offering a model, which parallels to what is now known in modern science as the TOE.

He theorized that the physical world is evolving and journeying towards the broadening of humanity's consciousness, the Noosphere, a term he borrowed from the Russian Mineralogist and Geochemist Vladimir Ivanovic Vernadsky (1863-1945). But he originally visualized the concept of Omega Point, which, according to him, is the zenith and peak of Consciousness. In his model, he categorized the Cosmos into four ascending realms, namely: physical, biological, our birthing us humans or what he calls the realm of hominization, and, finally, the realm of divinization.

This theory was considered radical and revolutionary during his time that his work, "The Phenomenon of Man," was only published after his death. But he must have been propitiated by the gods and goddesses for his contributions are now cited by Pope Francis in his latest encyclical "Laudato si." We are also told that way back in 2009, Pope Benedict XVI, Francis' predecessor, through his spokesman Federico Lombardo, was also quoted as saying that Teilhard was a heterodox author whose writings should now be studied.

It is this attempt of Teilhard at linking the physical and the metaphysical that inspired me to construct my own model.

As a start, the title of this chapter is based on a biblical verse taken from the King James Version, which uses "mansions" instead of "rooms" to describe the Father's house. This particular scriptural version was chosen because I believe the imagery better captures the infinite expanse of our Cosmos. People give different interpretations to the term "cosmos" (lowercase "c").

For instance, it is used synonymously with the terms *Universe, universe, multiverse, mega-verse, meta-verse, parallel universes, bubble universes, world,* and *nature* or *Nature.*

Etymologically, the term "cosmos" comes from the Greek word *kosmos,* which connotes a sense of order, harmony, and beauty. Order and beauty were first applied to things like clothing and jewelry. This is the reason why the term "cosmology" is also associated with the terms cosmetology, cosmetics, and cosmopolitan.

I used the term Cosmos (capital "C") to distinguish it from all these terms, which are only different parts or localized versions of the totality of the Cosmos (Edward Harrison, 2014).

Beyond our dense physical realm lie domains that in many respects are still unknown to us. These domains are not detectable by our ordinary senses yet they're out there. Thus, while I take the materialist perspective, I cannot ignore, much less dismiss, the existence of realms subtler than the physical. As Fred Alan Wolf (2001) remarks:

> *Some of us, including many scientists, don't agree with the objective of new materialism. We believe in our hearts of hearts, as*

*did the alchemists that came before us, that something far richer than materialism is responsible for the universe.*

Quantum physicists, cosmologists, and astrophysicists maintain that there could be an infinite number of domains out there. We are neither aware of them nor disturbed by their presence because they live in a dimension with frequency and vibration much higher and subtler than what we presently know. It could be that we're sharing the same space with them passing right around or through us, or we through them. Though these realms are separate, there's continuing communication and interaction between them. On our part, this is possible because we have the needed technology as well as the abilities to do so.

The efforts of physicists to discover the ultimate reality that could explain everything there is about the various dimensions of the Cosmos led to the development of the various field theories. These are the: (1) *Zero Point Energy Field* (ZPE) theory; (2) *Materialist Field* theory that expressed the materialist-rationalist school of the Greek philosophers, most notably, Lucretius, Democritus of Abdera, Plato, and Socrates, among others; (3) *Classical Field* theory, which is reflected in the mechanistic and deterministic views of Isaac Newton; (4) *Gravitational Field* theory that emerged from the General Relativity Theory of Albert Einstein; (5) *Quantum Field* theory introduced by Albert Einstein but pursued by Paul Dirac and the rest of the giants of quantum physics; and the (6) *Holographic Field* theory of the Cosmos that emerged from the *Superstring (or String, for short) Field* theory and its extension the *M Field* theory.

Their theories laid the foundation of the Cosmic Energy Field theory (CEF) I am advancing in this book as ultimate fabric of the Cosmos.

## The Zero-Point Field Theory

The ZPE was introduced by Max Planck in 1911 and considered to be the lowest state of energy level that fills up the void of space. This primal state, however, is not empty, inert, and passive. In modern science, there is no such thing as "a state of absolute nothingness" for this nothingness is crawling with elements so small and emitting tiny vibrations or quantum fluctuations that they cannot be detected by our standard physical measurements (Ray Fleming, 2012).

Albert Einstein shared in coining the term "zero-point energy." In his view, in the beginning was energy, manifesting itself in different forms like heat, kinetic, potential, nuclear, chemical, mass, and electromagnetic energy; and, then, this energy transformed itself into matter. What gives this energy its form is important for, as physicists maintain, without this "something," there would not have been any stars, planets, life, and humans existing today.

Quantum physicists describe the ultimate reality as a filament of vibrating energy often represented as a string. It is this primal element that animates and sets everything in motion. Undetectable by our ordinary physical senses, it is the subtlest state of existence, the realm of the invisible. Some scientists describe this primal state as one of Oneness since, according to them, at this initial state there is no distinction and differentiation. It is probably because of this that others refer to this state as the "Unified Field."

There are some parallelisms between the ZPE field theory and the *Divine Field Theory* (DFT). I use DFT to refer to the ancient creation stories about how the Cosmos started. These stories are handed down to us in the form of myths, legends, symbols, imagery, and metaphors. They implicate the existence of supernatural beings—gods and goddesses—who created everything and continue to intervene in the operations of the Cosmos.

The earliest known texts that document the history of the Cosmos can be traced in ancient Mesopotamia, particularly, in the cities of Sumer, Akkad, Babylon, and Assyria. Their records narrate the genealogies of the gods and goddesses, the beginnings of life, and the appearance of humanity, codified and sculpted in clay tablets.

First, dated to be written in 8,500 B.C., they are acknowledged to be much older by at least a thousand years than the Hebrew Bible's Book of Genesis. Many scholars now agree that what was written about the Cosmos in these ancient texts influenced ancient civilizations in Greece, India, Egypt, and Rome, including those of our modern times.

Ancient religions and beliefs about the primal order of the Cosmos are clear and straightforward. The primal state was one of disorder and chaos that has the potential to manifest itself in several forms. At first there was nothing, no form, only the void and darkness, but it was not inert only because the Spirit of the Lord hovered over it, holding back its potential to manifest. In the biblical account, for example, the Book of Genesis speaks of water that represented chaos, nothingness, and darkness:

> *In the beginning, when God began to create the heavens and the earth, the earth had no form and was void; darkness was over the deep and the Spirit of God hovered over the waters.*

The Chinese sage Lao Tsu portrayed the creation of our Cosmos as a confused spiral of primal elements depicted as an egg. Inside this egg were the two opposites, namely, the *yin*, representing darkness, passivity, and weakness" and the *yang*, representing "brightness, activity, and strength" (Hua-Ching Ni, 2003:14). This description is also reflected in the many writings of great philosophers several centuries later. For example, the Roman poet Publius Ovidius Naso or Ovid spoke of the same chaotic and confused state of the Cosmos (Gleiser, 1997:14-15):

> *Before the creation of the earth and ocean and sky,*
> *Alike was the face of nature in all here course,*
> *One indistinct chaos: a rough, disorderly mass*
> *Of inharmonious atoms confusedly mixed*
> *And lacking in all but lifeless and motionless weight.*
> *As yet no luminous sun enlightened the world.*

At this original state, there were no forms, no subject and object. There was only oneness. As the Chinese sage Lao Tsu (Hua-Ching Ni, 2003:79) expressed it:

> *When the subtle Way of the universe is all pervading, there is no longer any distinction between subject and object, between spiritual and material, between holy and unholy. All energies merge into harmonious Oneness.*

Sudhakar S. Dishit (1989) expressed this ancient belief of oneness in diversity in the primal state as:

*... the source of the entire cosmos and all cosmic activities relating to the emergence, existence and dissolution of the terrestrial phenomena that form the cosmic rhythm. And this ultimate reality is One absolute and indeterminable.*

In summary, the ZPE field theory speaks of a state of existence that began even before the singularity erupted into a fiery Big Bang. It speaks of a state before the beginning of time, space, and matter.

Our ancient ancestors symbolized this state nothingness as water over which the Spirit of the Lord hovered or the egg inside which are the two opposites, the yin and the yang.

## The Materialist Field Theory

Ancient Greek philosophers veered away from the metaphysical and god-centered view of our ancient ancestors by offering a materialist description of the origin and motion of the stars, moons, planets, and other celestial bodies based on the rigor of reason and philosophy. According to this view, everything in the Cosmos is made up of *material elements*, the equivalent of Aristotle's "*physis*."

But they differed as to what constituted the primal material particle. For Thales, the primary material of all things was water. Anaximenes considered air as the principle of life from which all things come from. Heraclitus saw fire as the beginning of all things, while Empedocles combined all the elements of earth, air, fire, and water. Aristotle contributed another element, *aether* (from a Greek word for "blazing") as the special substance out of which the heavenly bodies are composed.

From this materialist perspective, Aristotle investigated the ultimate source and reality from which all things started. Nature, to him, consisted of simple elements that are capable of self-movement, thus, possessing the potential to bring about changes and transformation. He argued that the principal generator of these changes is really the soul, which he referred to as the *entelechy*. According to him, the soul is the real substance as well as the source of movement and the final cause that animates all cosmic elements.

The Greek philosophers were careful in referring this soul to God. Instead, Plato postulated the existence of a Demiurge, a Divine Craftsman and Creator who constructed the Cosmos according to a prepared model based on the raw materials already available then (R. E. Krebs 2003).

It was Leucippus of Miletus and his disciple Democritus who were most influential in explaining the ultimate reality. These two philosophers introduced the idea of atom and the void. They viewed atom as the smallest indivisible unit and the primordial building block of the Cosmos. They argued that in the beginning there was only atom (literally "not able to be cut") and the void and that these two are inseparable. For them, the void simply means the space in which the atom moves. According to Leucippus and Democritus, in the beginning there were an infinite number of atoms, differing in size and shape.

These atoms moved in the void where collisions between and among them took place; atoms of irregular shapes got entangled with one another, eventually forming groups of atoms. It was in this manner that the Cosmos was finally formed. As Leucippus explained (Danielson, 2000:23):

> *The worlds are formed when atoms fall into the void and are entangled with one another; and from their motion as they increase in bulk arises the substance of the stars. Creation is simply a necessary consequence of the contact and merger of atoms in the void of the contact and merger of atoms in the void, although may have been brought about by sheer luck and chance.*

At first, simple elements were produced, but as the coalescence became more and more intense, simplicity gave way to complexity and what was once one became diverse. The famous Roman poet Lucretius utilized the alphabet as a metaphor to explain how the combination and permutation of only 26 letters in the English language can produce volumes of literature. Similarly, it was a mere handful of atoms that produced everything in the Cosmos.

In the view of the ancient Greek atomists, the void was necessary; the term void, however, was taken simply as an empty space or a place for the atoms to move around (Danielson, 2000:26, 29). For as Zeno (during the mid-5th century B.C.), declared: "movement is impossible," in the absence of an empty space (Danielson, 2000:25). Aristotle expressed this oneness of place and motion in his concept of "locomotion" (Danielson, 2000:37). For the Greek atomists, the void is the fabric of the Cosmos. Without it, life would not have appeared and we would not have emerged and developed to what we are today.

Philosopher Gottfried Leibniz (1670) also echoed this ancient belief of unity at the level of the monads (equivalent to the Greek's conception of atoms) and that reality can only be found in the One single source (see also Rodney Brooks, 2010):

> *Reality cannot be found except in One single source, because of the interconnection of all things with one another. ... I do not conceive of any reality at all as without genuine unity. ... I maintain also that substances, whether material or immaterial, cannot be conceived in their bare essence without any activity, activity being of the essence of substance in general. ...*

Since there were no physical evidences that could prove the existence of atoms then, further explorations of atoms were simply ignored over thousands of years. Atoms were simply thought to be a product of pure speculation. Nevertheless, the materialist view of the Greeks remained unchallenged until the first half of the 20th century when atoms began to be smashed into pieces using high-density particle accelerators. Results of these laboratory experiments revealed that atoms are after all not indestructible or indivisible. Tiny particles smaller than atoms exist in abundance.

**The Classical Field Theory**

Earlier physicists discover that atoms were point particles or dots, appearing as tiny specks of solid elements and floating in a vast space of emptiness and nothingness. Through a long process of chance coalescence and combination, these tiny dots produced bigger dots that were at the beginning simple but later became more complex as evolution progressed. But this contention has been challenged.

Later scientists say that the fabric of the Cosmos is not dot-like but more akin to "lines of force" or simply "fields." Michael Faraday contended that space is not empty but saturated with electric and magnetic fields. According to him, these fields are present whenever and wherever there is mass. They are also "the source of the gravitational force" (Carl S. Heinrich, 2012). Thus, the idea of field began to be entertained as the fabric of the Cosmos.

James Clark Maxwell later reinforced this field concept in his two published papers, one "On Faradays Lines of Force" in 1855-56 and the other "On Physical Lines of Force" in 1861-62. His greatest contribution was in constructing an extremely classy mathematical theory demonstrating that "a wave of electromagnetic energy would spread out into space like a ripple on a pond, changing the nature of space itself." This idea led to the view that light is simply a tiny band in a vast spectrum of electromagnetic waves (Nancy Forbes and Basil Mahon, 2014).

In effect, Faraday and Maxwell integrated electricity, magnetism, and light into one coherent and consistent theory. The field became known as the fabric as well as the storehouse and transmitter of energy that "pervades the physical world, the electromagnetic field" (Forbes and Mahon, 2014). Thus, a new cosmic paradigm emerged, portraying the Cosmos as soaked in a field of energy. W. Thiring (in Capra, 1975:208) had this to say about the field:

> The field exists always and is everywhere; it can never be removed. It is the carrier of all material phenomena. It is the 'void' out of which the proton creates the pi-mesons. Being and fading of particles are merely forms of motion of the field.

Newton expanded the materialist view by symbolizing the Cosmos as a mechanical clock, governed by fixed laws that determine its movements and changes. According to him, the world is like a machine, regulated by physical forces, referring to the laws of inertia, momentum, and motion. According to him, together with the law of gravity, these forces became the basic laws of Nature that ensure stability and order of the entire solar system. Obeying the rules of these forces, the Cosmos works in mechanically predictable ways and with clockwork precision.

Later, physicists contended that these basic laws of physics operate also in non-physical realities like our thoughts and emotions. But while

Newton conceived of time and space as the fabric of the Cosmos, he argued that they are unrelated and unconnected to each other.

This view changed with the emergence of the gravitational field theory.

## The Gravitational Field Theory

Einstein re-introduced the field concepts of Faraday, James Clark Maxwell, and Sir Joseph John Thomson to the equations of his general theory of relativity. From his efforts, the term "field" became popularized and began to be a catchword.

It was variously described as classical scalar field, gravitational fields, electric field, magnetic field, electromagnetic field, or energy field, to name a few. In deference to the Newtonian view, Einstein argued that everything in the Cosmos are intimately connected and that time and space stand on an equal footing. According to him, time represents another dimension, now known as the special relativity theory.

In addition, Einstein viewed the force of gravity as simply the resulting force of the presence of the object's weights, sizes, and mass (Brian Greene, 2011). If the fabric can be pictured as a flat sheet of blanket spread throughout infinity, the crumpled portions represent the effects of the various particles and elements that lie on top of what is otherwise flat in shape while the curvatures created by these particles produce the gravity that accounts for the behavior of these particles (Lewis H. Ryder, 1996). Thus, space, time, and gravity became embedded in the fabric of the Cosmos.

This new conception became known as the Gravitational Field theory.

## The Quantum Field Theory

In 1905 Einstein combined Faraday's view of light as a stream of particles with Maxwell's view of light as a stream of waves. He theorized that light exhibits the properties of a wave but also behaves like particles. Because of this new interpretation, the effect of gravity is not only confined to the curvature of space-time but also light (Henrik Antoon Lorentz, 2009:29). This observation by Einstein paved the way to the birth of the popularly known "wave-particle duality" of atomic elements, which as we all know by now is the foundation of quantum physics (Rodney A. Brooks, 2010).

But at that time this wave-particle duality was still unpopular, even unacceptable since it appears confusing and paradoxical. It is paradoxical since, as T. Norsen (in Rodney Brooks, 2014) explained: "particles are, by definition, localized entities that follow definite trajectories while waves are not confined to any particular path or region of space."

Scientists, however, observe that, in the quantum world, particles can appear, disappear, vanish, and reappear at any moment in space and through time. An observed electron, for instance, can disappear and reappear elsewhere in another region of the space-time continuum in a matter of microseconds.

In another case, a neutron can decay into a proton, an electron, and a neutrino in a matter of about 930 seconds, while a *muon* decays into an electron and a couple of neutrinos with a lifetime of about 2.2 microseconds (Brian Hatfield, 2014). This appearance and disappearance of particles introduced the quantum field theory (QFT).

QFT also explains the paradox that exists in the wave-particle duality by contending that disparate behaviors of particles and waves can co-exist in a coherent manner. This means that particles and waves can simultaneously be both local and nonlocal (Anthony Duncan, 2012).

The foundations of QFT were laid down by Paul Dirac and Heisenberg (Franz Mandi and Graham Shaw, 2013). As of this writing, it became one of the most important tool for understanding and explaining the microscopic world (Itzykson and Zuber, 1980). It became accepted as the "true theory of nature" and, we might as well say, "the true fabric of the Cosmos" (Novel laureate Frank Wilcek (2015).

This is not yet the end of our discussion, for the above field theories were still unable to integrate the force of gravity into the equation.

Their efforts began to be rewarded since these led to the development of the Holographic theory that springs out as a result of String theory and its extension M theory, believed to be the likely candidate to unify all known particles and forces into one coherent theoretical framework (McMahon, 2008).

## The Holographic Theory of the Cosmos

The *holographic theory* of the universe was discovered in 1968 by Gabriel Veneziano and M. Suzuki when they, by pure luck, stumbled upon a formula written by Mathematician Leonard Euler. The formula exhibited the features of a string. However, it did not draw much attention then, much less acceptance because it entailed a definition of the Cosmos that required 26 space-time dimensions. At that time, this was unthinkable.

But it made its comeback in the mid-1980s by proposing that the ultimate foundations of the Cosmos are no longer point particles but one-dimensional filament of vibrating energy called string, the length of which is equivalent to that of the Planck scale, a hundred billion times smaller than one atomic nucleus. Whereas interactions of point-like particles do not happen, in string theory, interactions can happen. As the theory goes, a string can vibrate at varying levels of frequencies and vibrational patterns from one region of space to another and through time (David McMahon, 2008; Brian Greene, 1999).

Then came its extension, the *M theory*.

Physicists discovered that strings have companion elements called *branes* (or *membranes*), which scientists now consider to be on "an equal footing" with strings (Peter West, 2012). Said to abound in the Cosmos, they continually collide with each other puffing off light that signals the creation of new and parallel universes (Tom Siegfried, 2016).

Like all other subatomic particles, branes, in their continuing motion, tend to collide with each other, in the process either annihilating each other or creating new universes. These brane collisions may have produced a muffled sound, similar to the popping of a balloon or something akin to the sound produced by the meeting of the male and female waters in the ancient Sumerian creation belief.

According to this view, the Big Bang happened when two of these branes collided with each other, creating a new universe in the Cosmos. At the moment of the collision, the four physical forces—gravity, electromagnetic, weak, and strong nuclear forces—were said to have been already interacting with each other at varying degrees of intensity.

In fact, the M theory infers that from the very beginning, gravity must have already been embedded, together with the other forces of physics. Gravity from other universes must have dripped into our space-time dimension.

This idea is quite revolutionary. We are no longer living alone in the universe. There are, according to this view, not only one but several universes, a concept now known as *Multiverse*, to distinguish it from *Universe*. Physicists even maintain that there could be an infinite number of universes parallel to ours that could be inhabited by higher forms of intelligence with highly advanced civilizations.

The mathematical equations say there could be millions of them out there. Even at only 0.05 percent of one million, there are still 500 extraterrestrial planets in our vast universe with their own Solar system. This is no longer the stuff of science fiction, the likes of Buck Rogers, Star Trek, the Matrix, or Space Odyssey. Science is now directing its technologies to visiting these other universes, re-engineering our DNA, and restructuring our molecules to prepare us for extraterrestrial living.

This is why quantum physics is promising and inspiring. The National Aeronautics and Space Agency (NASA), the military, and the enterprising are now cashing in to explore the world out there, sending our astronauts and satellites to colonize and populate the universe. Again, we are all dealing with speculations and dreams. But it is because of these dreams of ours that we've gone so far ahead now.

Among scientists there is a strong belief that we can enter into the other realms through wormholes, black holes, dark matter, and dark energy. For these scientists, galactic travel is theoretically possible with the appropriate galactic ships, which are now in fact being produced, tested, and perfected.

First implied by Einstein's theory of general relativity, wormholes are like tunnels between two places in the Universe. In theory, if you fall on one side of a wormhole, you'd appear on the other side almost instantaneously, even if it happens to be on the exact opposite side of the Universe. Wormholes are entrances that allow us to travel from one country to another without the need for passports and visas. But, more than that, they're also portals between two or more universes.

As Carl Sagan once said, "You might emerge somewhere else in space, some when-else in time." If this is so, then, wormholes can also serve as tunnels or passages that connect the physical, mental, psychic, and Consciousness realms and through which we can travel from one dimension to another. M theory's explanation of how our universe came to be is still new.

Nonetheless, it is very insightful, since it takes us back to the moment before the fiery Big Bang and allows us to explore other dimensions of the Cosmos. M theory points to an idea that our universe is just one bubble among other universes. Our universe is just one membrane within a much larger membrane constituting the entire Cosmos.

Physicists liken the entire Cosmos to an onion consisting of several layers lying on top of each other or to a loaf of bread sliced into several pieces, each of which represents one universe. If this is indeed true, it suggests that there are indeed other realms beyond our physical dimension of time, space, matter, and energy.

## The Cosmic Energy Field

I am introducing an alternative paradigm I call the "Cosmic Energy Field Theory". The CEF is a proposed model that seeks to link and integrate all the field theories into one whole unified field theory. It is a product of the developments that emerged from the various field theories, starting from the time of the ancient Greek philosophers, to that of Newton's, Einstein's, and the founders of quantum physics. It serves as the field on which everything happening in the Cosmos is played out. It symbolizes the ultimate fabric of the Cosmos.

But it is not totally void or absolutely empty for it is teeming with virtual elements ready to materialize themselves as solid particles in an infinite number of forms. At its initial condition, everything is in a state of potentialities and possibilities. Once these potentialities materialized, they immediately sprung to life giving rise to diverse forms and structures from simple to complex.

It is on the CEF that everything appears out of nowhere and also vanishes into nowhere.

It houses the ultimate source of energy that gives form, sustains life, and the principle that governs the actions and behavior of the four cosmic realms as one whole unit. This energy manifests itself differently in the various realms. In the physical level, the four forces of nature direct the inhabitants' behavior and interactions, namely, gravity electromagnetism, weak nuclear force, and strong nuclear force.

Gravity, coming as it was from the other universe parallel to ours as a result of a collision, operates at the large-scale structures and assigns all gigantic structures like stars, planets, moons, etc. to their proper stations. The last three forces operate at the quantum level and it is partly from the interaction of these forces that give rise to order, harmony, and beauty we see and witness today. These energies resonate across the other dimensions. Information triggered in one dimension is instantaneously relayed through all the other and back in a continuing process. This makes the Cosmos a living organism.

The act of categorizing and classifying the CEF into various fields does not imply separateness to the whole unit we call the Cosmos. Recent findings provide convincing evidence that all the cosmic fields are interconnected and interdependent at their most fundamental level. As Lynne McTaggart (2007) states: "The existence of the Field implies that all matter in the universe is connected on the subatomic level through a constant dance of quantum energy exchange."

Another feature of the CEF model is that it is able to store memories of past particle interactions. Physicists theorize that this is due to the interactions of matter by gauge bosons such as photons for the electro-magnetic forces (Ray Fleming, 2012). Gauge boson is considered the last particle in the Standard Model that is believed to possess the ability to store information, ever ready to be recalled from memory and even accessed whenever one particle interacts with it. As Seymour (1992:61) explained: "If two particles have interacted in the past, then each particle carries a memory of that interaction which can be simultaneously recalled, so subsequent measurements on the pair will always be correlated."

All the information and events happening in the entire Cosmos are recorded in this particle. We can refer to this as the "cosmic data bank" or what the mystics call Akashic Record.

# 2

# And the Cosmos Dances to the Tune Played by the Pied Paper

*Everything is determined, the beginning as well as the end, by forces over which we have no control. It is determined for the insect, as well as for the star. Human beings, vegetables, or cosmic dust, we all dance to a mysterious tune, intoned in the distance by an invisible piper.* - Albert Einstein

*It is not that the theory of quantum mechanics is a strange description of Nature but that Nature herself behaves in a surprising and counter-intuitive way.* - Jim Al-Khalili

*The pursuit of nature's smallest parts is one of mankind's oldest quests... Each step deeper into the Micro-World has revealed surprises, and expanded our understanding of nature on all size scales.* - Robert L. Piccioni

*Mysteries always appear and they never fail to put us in awe of our history and our heritage. The universe has never failed to puzzle our ancestors and even today, after thousands of years, it still goes undefeated.* - Vedang Sati

*In the world of the very small, where particle and wave aspects of reality are equally significant, things do not behave in any way that we can understand from our experience of the everyday world...all pictures are false, and there is no physical analogy we can make to understand what goes on inside atoms. Atoms behave like atoms, nothing else.* -  John Gribbin

*The more we delve into quantum mechanics the stranger the world becomes; appreciating this strangeness of the world, whilst still operating in that which you now consider reality, will be the foundation for shifting the current trajectory of your life from ordinary to extraordinary. It is the Tao of mixing this cosmic weirdness with the practical and physical, which will allow you to move, moment by moment, through parallel worlds to achieve your dreams.* – Kevin Michel

*All the wonders of quantum physics were learned basically from looking at atom-smasher technology. ... But let me let you in on a secret: We physicists are not driven to do this because of better color television. ... That's a spin-off. We do this because we want to understand our role and our place in the universe.* – Michio Kaku

Could scientists have indeed trespassed into the world of metaphysics? If this is so, can quantum physics discover a new model that could explain the metaphysical world as well as connect both the physical and metaphysical realms as one integral unit?

As I dive deeper into the quantum world, the more I realize that I am drawn into the metaphysical and the spiritual, into the world of subjectivity and speculation. I've crossed the threshold of quantum physics and ventured into what Erwin Schrödinger called the metaphysical zone. I know I am not alone. The founding fathers have accepted that they must have already stumbled into something beyond the horizon of physics and as a result encroached on a nonphysical world, whose significance they acknowledged.

In the words of Erwin Schrödinger (1964:118), metaphysics is "the vanguard, establishing the forward outposts in an unknown hostile territory" and that "we cannot do without such outposts" for "it is the scaffolding, without which further construction is impossible." According to Schrödinger, eliminating metaphysics only means "taking the soul out of *both* art *and* sciences, turning them into skeletons incapable of any further development." "As we go forward on the road of knowledge," he continued, "we have got to let ourselves be guided by the invisible hand of metaphysics reaching out to us from the mist …"

Lynne McTaggart (2009) remarked that deep within the founding fathers had some inkling that they had already trespassed into the mystical zone. She noted that on their own personal initiatives, the founders went to the extent of studying the world's ancient religions.

Indeed, Pauli examined the Qabbalah, Bohr the Tao and Chinese philosophy, Schrödinger, the Hindu philosophy, and Heisenberg, the Platonic theory of ancient Greece. Bohm was identifying his 'implicate order' with Eastern spirituality (John Luckey, 2015).

If this is so, this could pave the way to linking both the physical and metaphysical realms since quantum physics could provide, according to Taggart, the scientific substantiation of "areas which have largely been the domain of religion, mysticism, alternative medicine or New Age speculation." Cox and Forshaw (2012) were even more specific by including the esoteric world as one realm in the study of metaphysics. To quote them:

> *Quantum theory is perhaps the prime example of the infinitely esoteric becoming the profoundly useful. Esoteric, because it describes a world in which a particle really can be in several places at once and moves from one place to another exploring the entire Universe simultaneously.*

With this new worldview, a path is laid for quantum physics to enter the realm of subjectivity, metaphysics, and the spiritual in spite of its materialist orientation. This view is in fact raised in several discussions in the past. On the issue of subjectivity and objectivity, the giants of physics acknowledge that at the quantum level, this distinction disappears. As Physicist Freeman Dyson (1979) argued:

> *When we are dealing with things as small as atoms and electrons, the observer or experimenter cannot be excluded from the description of nature. On the level of subatomic physics, the observer is inextricably involved in the definition of the objects of his observations.*

Werner Heisenberg (1958:88) also maintained that: "today 'simultaneity' contains a subjective element." In this respect, Ilya Prirogine (1997 opined that "the role of the observer … gave quantum mechanics its subjective flavor."

On the subject of metaphysics, quantum physics found out that entering the world of atoms is really entering the world of the nonphysical since atomic elements no longer possess mass, position, location, size, weight, and no longer subject to the restrictions imposed by our body and our four dimensional realm of space-time.

On the issue of spirituality, Nobel Laureate Brian Josephson of Cambridge University-England advanced that: "Once you begin to ask questions about states of consciousness, you reach a region where thinking in spiritual terms is a possibility" (in Gail Bernice Holland, 1998:14). Some quantum physicists also entertain the idea that Consciousness could be the "Soul."

While this idea has not been further pursued, Wolf (1999, 2000) maintained that this could not just be summarily dismissed: "I agree that science, at first glance, may appear to be the worst place to go for answers about the soul. But we shouldn't be too hasty in rejecting it. Deep down at the atomic level, we only see particles, waves, and energy, and if the Soul's existence can be entertained at all, then, it must lie beneath these subatomic elements."

Wolf, of course, is expounding on the materialist stance of quantum physics, since the soul remains an emergent property of the physical brain. Theoretical Physicist Amit Goswami (2001, 1993) is more emphatic when he said that this materialist and evolutionist view of the existence of the soul stems from the conventional assumption that matter, waves, and energy are the building blocks of all things, including consciousness and the soul.

But what explains the reluctance of quantum physicists to go into the metaphysical?

Schrödinger says that metaphysics uses a deductive method, which is the opposite of science's inductive approach that requires rigorous experimentation of empirical facts. He argues that it is metaphysics that may have turned into physics and if ever physics enters into the realm of metaphysics, it should not be the *a priori* kind suggested by Immanuel Kant, a view, which mainstream thinking is not yet ready to take.

Even if it ventures to tread this path today, Edgar L. Owen (2013) says it's still most unlikely that physicists would go deeper into this subject because of its strong adherence to mathematics. But, from its perspective, reluctance does not mean denial of the existence of the metaphysical realm. It simply means that the world of metaphysics is not within its domain.

I am mindful of Fred Alan Wolf's exhortation to me that the findings of my exploration are not intended to serve as conclusions derived from my "research into quantum physics." Rather, they serve as a reminder to me that there, too, are many quantum phenomena in the metaphysical world expressed in Religion and Mysticism thousands of years ago before quantum physics was born.

## The Story of the Atom

The quantum world is the realm of the small and the smallest, the tiny and the tiniest you can possibly imagine. While we cannot see atoms with our naked eyes, physicists know that they are there, observable as blinkering lights in a "now-you-see-now-you-don't" pattern.

Atoms were among the first to be discovered as the accepted fundamental elements of nature. The Greek scientists Democritus and Leucippus thought of them as indivisible and indestructible. But today we learn that once subjected to high-intensity energies and made to collide at speeds close to that of light, atoms can be smashed into smaller pieces.

At the beginning of the 20th century, physicists still doubted the existence of atoms because these could not be observed in the laboratory. The nucleus of an atom was discovered only in 1911, proton in 1919, and neutron in 1932. Upon their discovery, estimates of their sizes, weights, and lifespan were made. The atom is less than two billionths of an inch in diameter, so tiny that it is almost inconceivable. Yet, it is considered huge compared to the nucleus, proton, neutron, and electron. The nucleus is about 100,000 times smaller than the whole atom. The neutron is about 1,839 times larger than the electron and 12 times smaller than a proton.

We are told that the world of an atom is so compact, yet, it is populated by hundreds—maybe even thousands—of subatomic particles, many of which might still be waiting to be discovered. What is even mind-bending is that the atom is 99 percent space, leaving only one percent for its populace (Ali Sundermier, 2016). To measure these particles is quite a formidable challenge.

Due to the short distance allowed by their confinement, electrons move with estimated velocities of 600 miles per second. But this is almost nothing compared with the velocities of the elements inside the nucleus, which are estimated to travel at about 40,000 miles per second (Capra, 2000:62). Nevertheless, scientists are able to estimate the sizes and weights of these particles (see Krauskopf and Beiser, 1991:143; John Gribbin, 1984:31).

If in our wildest dreams we are able to ride one of the electrons, we won't feel anything moving despite its dizzying speed. From our relative position, it will be the nucleus moving around the electron. Heinz R. Pagels (2012) expressed that if one could ride the waves of light, the way a pilot maneuvers a plane to ride on a stream of air 30,000 miles in the sky, one would not feel the light wave moving, nor sense that one is travelling at a speed of 186,000 miles per second.

But why don't they cross into each other's path? What brought about order and harmony? Are there laws governing the Cosmos that move its constituents to dance to their tunes in perfect balance and symmetry, as the Pied Piper did to the mice and children?

For those unfamiliar with the tale, the metaphor is set in 1284 in the town of Hamelin, Lower Saxony, Germany. This town was facing a rat infestation, and a piper, dressed in a coat of many colors, appeared. The piper promised to get rid of the mice and rats in return for a payment and to which the townspeople agreed. Sounding his flute, the piper got rid of the rodents by luring them away from their holes and led them into a river where they dove and drowned.

But the town dwellers refused to pay him. The furious piper left, vowing revenge. On the 26th of July of that same year, the piper returned. But this time the children, with their parents unaware of what was going on, followed him into a cave far, faraway. In spite of the peoples' earnest efforts to know their whereabouts, the place where they went, disappeared, and was never located, the children were never seen again.

The world of an atom is closely similar, only it's not dealing with rodents, mice, and children, but electrons, protons, and neutrons. So, what tune was the Pied Piper playing? The classical theory of Newton and the relativistic theory of Einstein were able to approximate how order was achieved at the level of large-scale structures.

According to their views, the law of gravity is responsible for holding heavy objects together, thus, allowing all the constituents of our solar system to be in their proper stations and orbits in perfect harmony. Without gravity, every celestial body would have exploded, disintegrated, and flown farther and farther outward into space and we would not have fixed our feet on the ground. Gravity serves as the harmonizing and equilibrating tune that choreographed the harmony we witness in our celestial bodies.

At the atomic level, there are forces that make it sure that the electrons never collide with the nucleus by pushing all other particles that dare go near the protons and the electrons. Three forces, also called interactions, have been identified: the electromagnetic, strong nuclear, and weak nuclear forces. At its highest level, *electromagnetism* holds the atoms and the molecules together. At the lower level, it binds the negatively charged electron and keeps itself in orbit around the positively charged nucleus. In cases where there are more than one electron orbiting an atom, the electromagnetic force prevents these electrons from colliding with each other.

But as one goes down to the lower level, the effect of *electromagnetism* can no longer be detected. Another force, known as the *strong nuclear force*, comes into play, holding the protons and neutrons together (Robert E. Krebs, 2003:30; John Scalzi, 2003). Gamow (1972) argued that this force is not purely electrical since the neutron does not carry any electric charge at all.

At the lowest level is the *weak nuclear force*, so called because it cannot hold the nucleus together. Its range of interaction is shorter than the nuclear force, but found to have influence in the interactions of leptons (Klauskopf and Beiser, 1991:252; Capra, 1975:215).

Just like a rotating electric fan, the rotating blades represent the electrons orbiting around the nucleus. An electron is situated in a particular region where it continually orbits an atom's nucleus. There can be more than one electron orbiting a nucleus. Each electron maintains its own state and position. One electron may be located higher or lower than the other. Though it may be analogously incorrect, one cannot help but think of the atom's organization similar with our solar system.

The sun as the nucleus is at the center, while the planets encircle around it the electrons. Because of the forces governing the movements of the planets, no collision is possible, at least for the next billions of years, when our sun self-destructs.

All around our solar system is an immense expanse of almost limitless space, populated by a bigger system of galaxies and constellations that are, in turn, also populated by diverse cosmic bodies and entities. Russian-American Physicist George Gamow reinforced this atom-solar system analogy by comparing their masses, citing that "the atomic nucleus contains 99.97 percent of the atomic mass as compared with 99.87 percent of the solar system (1972:136).

We cannot feel the lightning speed at the macro level. The Earth completes its revolution in 365 Earth days; Venus in 225 Earth days; Mercury in 88 Earth days; Mars in 1.8807 Earth years; Jupiter in 11.8565 Earth years; Saturn in 29.4 Earth years; Uranus in 84.02 Earth years; Neptune in 164.79 Earth years; and Pluto in 247.32 Earth years.

Planets move so slowly that we don't even feel the Earth moving at all as it orbits the sun. The only thing we observe is the sun rising in the morning and setting in the evening. In fact, for thousands of years, our ancestors believed that it is the Sun moving around us, a theory proven to be untrue some 500 years back only.

While the three forces above explain a lot about the order and harmony happening in the world of atoms and their sub-particles, quantum physics also discovered several happenings at the quantum level that were considered perplexing because they cannot be fully explained by the laws of logic, physics, and causality.

## The Existence of Opposites

In 1928, Paul Dirac already theorized that energy can exist in two different states—positive and negative—but are identical to each other in both size and mass. He was proven right. In the 1930s, physicists discovered that particles have their opposites. The first to be discovered was in 1932 when American Physicist Carl Anderson noticed a twin particle called *positron,* which is the positive state of electron.

In 1934, Fermi broached the idea of a *neutrino,* Italian for "little neutron," that has no electric charge, no mass, neutral, no tendency to interact with matter, but capable of knocking out protons out of nuclei. He was also proven correct. In 1956, a team of American physicists led by Frederick Reines and Clyde Cowan, using a fission reactor, detected *antineutrinos* in their laboratory, namely, *antiprotons, antineutrons,* and *positrons.* Together with their opposites, these particles began to be known as the first generation of *leptons* (Davies and Gribbin, 1992:153).

Anderson discovered two other particles belonging to the second generation of the lepton family, namely: *muon* (later classified as meson) and *antimuon. Muon* is the same particle that Japanese physicist Hideki Yukawa also discovered and later confirmed by British Physicist Patrick Blackett (Percy Seymour, 1992:58). In 1947, British physicist Cecil Frank Powell discovered another particle the *pi meson.* In 1975, the American physicist Martin Perl discovered an electron-like particle he called *tau lepton,* or *tauon,* considered to belong to the third generation.

Thus, today, physicists speak of different "flavors" under each of these twin particles. For example, the three "flavors" of leptons are the electron and the electron neutrino, the *muon* and the *muon neutrino,* and the *tauon* and the *tauon neutrino.* There are also three flavors of antileptons: the antielectron (positron) and the electron antineutrino; *theimuon* and the *muon antineutrino;* and the *tauon* and the *tauon antineutrino.*

All together, there are 12 leptons and antileptons, considered as fundamental particles since they are indivisible (as of this writing), in spite of the discovery in 2012 of the Higgs boson, believed by many to be the God particle. Franz Mandi and Graham Shaw (2013) explained this as a composite field: "The electrons and positrons themselves can be thought of as the quanta of an electron-positron field." Since then, physicists have concluded that every particle or matter has its equivalent anti-particle or anti-matter (Gary Zukav, 2001:235; Iain Nicolson, 2007).

Opposites also exist in the macro world. Omraam Mikhaël Aïvanhov (2011) in his book titled *Cosmic Balance: The Secret of Polarity*, observed that creation is the manifestation of polarities—positive and negative, day and night, light and darkness, masculine and feminine, active and passive, life and death, or movement and stagnation. He noted that creation has been a continuing act of successive divisions.

Quoting the Bible, he noted that on the first day, light was separated from darkness; on the second day, the waters on high from the waters below; on the third day, the waters from the dry land (Genesis 1:4-10). He also cited that in biology "a cell, the smallest element of a living organism, reproduces itself by division, by splitting in two."

Omraam was also quick to say that there are no moral issues involved in these polarities; the word "positive" does not connote "something good and constructive" nor does the term "negative" suggests "something bad and destructive." William Shakespeare (Hamlet, II, ii, 249) says it's our thinking that makes it so: "There is nothing either good or, bad, but thinking makes it so."

But how do we explain the birth of something from nothing? According to quantum physics, when two opposites are released in the same direction at a speed close to that of light, they annihilate and vanquish each other at the point of contact at the other end. Physicists argue that matter must have outnumbered antimatter from the very beginning by a ratio of a billion and one particles for every billion anti-particles of the same kind. When the annihilation of particles and anti-particles took place, the surplus particles—protons, neutrons, and electrons—remained (Paul Davies, 1983:29-30).

Cosmologists as well as astrophysicists are convinced that this imbalance in the number of opposites was created from the primeval Big Bang otherwise, as British Cosmologist Paul Davies remarked, there would have been nothing left. There's also another ticklish issue. Why should there be something instead of nothing?

In spite of this tug-of-war relationship between the two opposites, quantum physicists discovered a basic symmetry that emerged in the interaction of opposites. They maintained that while opposites vanquish each other, they also exhibit perfect complementarity.

**The Principle of Complementarity**

Danish Physicist Neils Bohr (1958:57) advanced the "principle of complementarity" in 1927. He saw particle opposites as complementary representations of the same reality, exhibiting an intimate relationship and interconnectedness at the tiniest level of existence. Eugene Wigner (in Wolf, 1981:139) emphasized this view when he said that: "The most general physical properties of any system must be expressed in terms of the complementary sets of that system."

Though particles appear in contradictory ways, according to this principle, reality cannot be complete without the opposites and one cannot fully understand reality without understanding its opposite. Opposites are not isolated or separate, but, as in a jigsaw puzzle, they form complementary pieces of the same reality. Michio Kaku (1998:4) expressed that: "One of the most remarkable features of Nature is that its basic laws have great unity and symmetry when expressed in terms of group theory."

This principle extends to the macro level. The Cosmos manifests itself in opposites and while opposites are innate in creation, there is, at the same time, equilibrium, symmetry, and complementarity, a cosmic principle, which Fritjof Capra (1975:124) interpreted as the "metaphysics of quantum theory."

Omraam narrated the story of how God created the Cosmos in the first book of the Zohar. According to the legend, before everything came to be, there was the Holy One, the unity of the One, or *Aleph*, which is attributed to as the first letter of the alphabet (*Alpha*). In order to make Himself visible to the physical world, the Holy One has to become two, also attributed to the second letter of the alphabet *Beth* (or *Beta*). In creating the Cosmos, the Holy One had to manifest himself outside in opposites and thus become two.

The Book of Concealment in Zohar is quoted to begin with these words: "The Book of Concealment is the book of the equilibrium of balance" and that "the symbol of balance or scales dominates the whole of creation" (Omraam, 2011). In the world of the divine, while everything in creation is governed by opposites and dualities, harmony is possible if one is conscious of the "One." For, as Omraam observed, there is only one unique reality but manifested to us as polarity. It is from this polarization that diverse forms and structures emerge and interact in great harmony and order as one coherent unit.

The Rgveda, which is dated 5,000 years ago, already noted that while the physical Cosmos is marked by polarity, at its highest realm it has oneness and unity. Consciousness transformed itself into opposites that make it possible for the opposites to be capable of self-consciousness.

## The Cyclical Process of Creation and Annihilation

The Big Bang theory can now be replicated inside huge high-speed particle accelerators or atom smashers. These machines are circular in shape extending for some miles in circumference under the ground where subatomic particles are released in opposite directions and made to collide with each other at velocities close to the speed of light.

What has been observed in these controlled experiments is that if a particle meets an anti-particle, they cancel out each other and vanish without traces. But in the course of the annihilation, new higher energy particles identified as *photons* are created at the point of contact. A neutron disintegrates into several pieces spontaneously, which is a process called "beta decay" and is transformed into a proton, accompanied by the creation of an electron and a new massless particle called *neutrino*.

The proton, the electron, the photon, and the *neutrino* are considered stable particles in the sense that they live forever, except when they are put also in a collision course. All the other particles (at least those discovered until the mid-1970s) are considered unstable, since they undergo a process of decay almost immediately after they are created. These unstable particles disappear after a few "particle second" and become invisible but they leave tracks at the collision point in the bubble chamber (Capra, 1975:187, 217-225).

In any case, at the moment of collision, a cyclical process of annihilation and creation takes place. Gary Zukav (2001:216) made this observation: "What we have been calling matter constantly is being created, annihilated, and created again. This happens as particles interact and it also happens, literally, out of nowhere."

German Physicist Hermann von Helmholtz hinted that this is also applicable at the macro level. By regarding the Cosmos as a closed and isolated system, i.e., there is nothing outside of it, the entire universe is inevitably headed towards a "heat death," ending in a Big Crunch or Big Freeze. Alex Vilenkin expressed the view that this Big Bang-Big Freeze/Big Crunch phenomenon marks the cyclical process of birth and death of the Cosmos. In his words (2006:22):

> *At the initial moment, which we now call the big bang, all matter in the universe is packed into a single point, so the density of matter is infinite. The density decreases as the universe expands and then grows as it recontracts, to become infinite again at the moment of the "big crunch," when the universe shrinks back to a point. The big bang and the big crunch mark the beginning and the end of the universe.*

This cyclical behavior of the Cosmos has long been observed thousands of years ago in the beliefs of our ancient ancestors. H. Zimmer (1972) cites the Hindu view of a God whose bodily movements signify either creation or annihilation:

> *His gestures wild and full of grace, precipitate the cosmic illusion; his flying arms and legs and the swaying of his torso produce—indeed, they are—the continuous creation-destruction of the universe, death exactly balancing birth, annihilation the end of every coming-forth.*

Joseph Campbell (19724:63) was more poetic in describing this cyclical process of creation and annihilation by referring to the Hindi belief:

> *Brahma sits on a lotus, the symbol of divine energy and divine grace. The lotus grows from the navel of Vishnu, who is the sleeping*

*god, whose dream is the universe. . . . Brahma opens his eyes and a world comes into being . . . Brahma closes his eyes, and a world goes out of being.*

Other prehistoric beliefs applied this concept to humanity and civilization as follows:

*Through energy the worlds were made, and through that energy they make progression; through energy the forms unfold and die; through energy the kingdoms manifest and disappear below the threshold of the world which ever is, and which will be forever* (Alice A. Bailey, 1993:556).

Omraam (2011) was also able to give a parallel description of this cyclical process of creation and annihilation employing the analogy of the biblical grain of wheat: "Unless a grain of wheat falls into the ground and dies, it remains a single grain; but if it dies, it bears much fruit" (John 12:24). According to him, a seed has to split itself into two parts for a new life to sprout and grow. What was once one becomes two to eventually give rise to three. As a new form of life emerges, then, the seed disappears and dies. Death plays a role in nature by giving birth to new elements. It is also in this manner that death and life becomes complementary. What comes after the flourishing of the sprout is the bearing of fruits and abundance.

**The Dual Nature of Atoms**

The dual nature of matter emerged from a long history of experiments that began towards the end of the 18th century. It was being debated in these experiments whether light was composed of a stream of particles or wave. Newton believed that light is a stream of *particles* or "corpuscles," not *waves*. The results of many studies, however, were mixed and divided. Nobody knew for certain what light was really all about.

In order to shed light on this dilemma, scientists, beginning in the late 19th century, started to make experiments using the light emitted from a *black body radiation*. It is called black because the body or object looks totally dark as it absorbs light that tumbles into it. In 1900, Max Planck theorized that the light wave emitted from the black body radiation is in the form of tiny packets of indivisible or *discrete units* of energy, called *quanta*. A packet of energy, as Percy Seymour further explained, was called "a quantum of radiation, and collections of such packets were called quanta" (1992:55).

According to Planck, discrete units are the smallest units (also termed as "Planck unit") that can be measured or quantified and one could not get any amount smaller than this (Luckey, 2015). The existence of discrete energy levels was, in fact, proven in 1914 by the Frank-Hertz experiment (Peleg et al., 2009).

In 1905, or five years after Planck's ingenious idea about light, Einstein in his paper on "Photoelectric Effect" integrated the two opposing views of quantization or measurability of light and advanced the idea that light is *both* particle *and* wave. According to this view, the particle carries with it an energy that keeps it streaming in continual motion. Waves behave as particles and particles behave as waves (see also Steven Holzner, 2013).

Einstein then proposed that light consisted of tiny burst of energy called *photons*, which serve as carriers of light. This idea led to the quantum theory of the nature and behavior of light as both *wave* and *particle*. Further experiments later revealed that these wavelike properties applied not only to light, but also to "a whole host of other experiments involving radiation—from radio waves to x-rays" (David Bohm, 2012).

The double-slit experiment has been used several times to demonstrate the dual nature of electrons. Thomas Young was considered the first to perform this now famous mind-boggling experiment that started between 1799 and 1801. As it is done today, a single electron is made to pass one after the other through two slits or small openings and end at a screen posted at some distance from the slits. What came out on the screen is a random spot of dot or point particle (Luckey, 2015).

But when the two slits are bombarded with a large number of electrons, their collective effect appears as "undulant double-slit" pattern, similar to a rolling wave in an ocean, suggesting that the wave "represents the state of random discontinuous motion of particles" (Gao, 2014).

I will continue to discuss these strange quantum discoveries below with an ending note that the Chinese sage Lao Tsu (2003:32-38, 137-139) already hinted this dual behavior of matter thousands of years ago. While physicists speak of matter, energy, and wave, Lao Tsu spoke of "materiality and spirituality" that is innate in each small particle, the "subtle essence of the universe," "so elusive and evasive," "unveiling itself as images and forms," yet so real.

**The Pilot Wave**

French Physicist Louis De Broglie theorized that there exists a second wave that determines the precise position of particles at any particular time. According to him, the wavelike behavior of a particle makes it possible to travel ahead of the particle and, in so doing, probe the conditions that lie ahead. This second wave is picturesquely described as the "pilot wave."

The data gathered from the survey are instantly relayed back and projected on the panel board of the plane, enabling the pilot to respond and adjust the plane's direction, speed, or altitude before any eventualities, otherwise unknown to the pilot, could occur.

According to de Broglie, the wavelike behavior of a particle makes it possible to travel at a speed much faster than light since wave is energy. It is its "phase velocity," not its solid particle that travels to explore the region of space ahead (Seymour, 1992:57). Thus, this theory suggests that the pilot wave does not conflict with the relativity theory of Einstein that sets limit to the travel of particles to the speed of light.

**The Observer Effect**

In its primal state, the wave nature of the electron goes its own endless motion and stays in motion until an observer comes in and disturbs it. When an electron senses that it is disturbed by the intrusion of an observer, its original wavelike behavior freezes and manifests itself as a particle at a particular region in space before the observer.

The process of observation triggers the collapse of an electron's wave function. The manifested reality exists only because of the intrusion of the observer, leading Freeman Dyson (1979:9) to remark that: "The effect of observation is absolutely fundamental to the reality that is revealed." Unless disturbed by the observer, the capability of an electron to transform as a solid particle remains in a state of potentiality and possibility.

Observation disrupts the electron's original state. As Wolf (1981:2) averred: "… every act of observation made of an atom by a physicist disturbed the atom… The meagerest attempt to observe an atom is so disruptive to the atom that it is not possible to even picture what an atom looks like."

One can, therefore, account for two realities or states. As Davies explained, the system is "capable of changing with time in two completely different ways: one when nobody is looking and one when it is being observed" (1988:168).

## The World of Potentialities

In its original state, the electron exists in all possible states. All observable phenomena possess the intrinsic potential to be actualized. But the form they will eventually take cannot be determined with accuracy. Anything one can imagine as their outcome is only a possibility without any assurance that they will unfold in the physical realm.

In 1920 Max Born theorized that the electron is really a measurable wave of probability, or "probability wave," which, according to Wolf makes both the particle and wave nature of an electron a probability function (Wolf, 1981:101-102, 105; see also Fritjof Capra (1977:56). At its original state, an electron can be theoretically located only at any given range of random positions around the nucleus. But because of its wave behavior, it can be anywhere within any of its orbital range.

## Quantum Jump

Physicists also observed that an electron can jump from one station to another depending on whether it absorbs or emits radiation. For example, from its lowest orbit, also called "ground state," an electron can jump to a higher orbit if it receives the needed energy. It can also go back to its original ground state if it gives off its surplus energy in the form of electromagnetic radiation (Percy Seymour, 1992:56; Capra, 2000:59-60).

This phenomenon is known as quantum jump since they leave no traces in between orbits. As Columbia Physicist I. I. Rabbi remarked (quoted in Cynthia Sue Larson, 2013):

> *The atom is in one state and moves to another, and you can't picture what is in between, so you call this a quantum jump. In quantum mechanics, you don't ask what's the intermediate state because there ain't no intermediate state. It passes from one to the other in God's mysterious way.*

Basil F. Hiley and David Peat (2012) writing in memory of David Bohm explained that in their conversations with Bohm, they all agreed with the idea that we do not know precisely what happens to subatomic particles in between quantum jumps. Technically, physicists described this simply as a statistical theory "in the sense that the description of the individual particle can only be given in terms of probability of it being observed at a certain point in space-time." According to them:

> *There is no description of the individual except in terms of its possible observation by some suitable measuring device... There is only a sequence of results of measurements, with no possibility of discussing what goes on between measurements.*

In the deterministic and causal world of classical physics, the notion of quantum jump can be immediately dismissed as inaccurate, illogical, and illusory. Classical physicists believe that there has got to be a solid theoretical foundation to explain this.

Albert Einstein and his colleagues Boris Podolsky and Nathan Rosen (known by their acronym EPR) tried to arrive at an explanation by performing thought experiments. The results of the experiments, however, consistently defied the classical view of cause and effect and the idea that no material particles can travel close to or faster than the speed of light, advanced in the special relativity theory of Albert Einstein.

Consequently, he made this now oft-quoted remark: "I cannot seriously believe in [quantum theory] because ... physics should represent a reality in time and space, free from spooky action at a distance."

A related concept that emerged from the dual behavior of particles is quantum tunneling and quantum teleportation.

**Quantum Tunneling**

Quantum tunneling was first introduced by F. Hund in 1927 and was originally known as "barrier penetration." It is a concept used to express the potentiality of particles to barrel or tunnel through any barrier and land on the other side, without changing or destroying the latter. Particles like *neutrinos*, *axions*, *photinos*, and *gravitinos* are high-density particles; they are so subtle and tiny that they can penetrate even a block of solid lead light-years thick.

Neutrinos can be detected as "whimpers" and are known to have no electric charge, no electromagnetic force, no size, no spatial extent, no radius, and may not possess a mass at all (Davies and Gribbin, 1992:176; Leon Lederman, 2012:344). This is the reason why these particles are also called "ghost particles.' Paul Davies and John Gribbin (1992:207) described this concept of tunneling more succinctly as follows:

> *Tunneling is the quantum mechanical process by which a particle can penetrate a classically forbidden region of space (for example, passing from two separate points A and B without passing through intermediate points). The phenomenon is so named because the particle, in traveling from A to B, creates a sort of "tunnel" for itself, bypassing the usual route.*

The reason why we cannot observe this in the macroscopic level is because large-scale objects have shorter wavelengths. But some researchers maintained that there is evidence that this tunneling in the world of small things is also happening in the world of big things. Larson (2012) raised this observation by citing some concrete human experiences:

> *Possibly you noticed a sock or two missing after you washed it, somewhere between the washing machine and the drying machine, most inexplicably leaving you one or more lone socks with no*

*mates... What's most curious is when the socks simply seem to vanish altogether, and all other possibilities have been investigated.*

*Perhaps you've had an experience where you are absolutely sure that you set down your wallet or keys in a certain place, only to return and not be able to find them there. You then search around the place a bit more unsuccessfully, and return again to that same place where you first looked...and there they are.*

This is the reason why scientists are now closely monitoring events happening at the micro level. As Larson (2013) observed:

*Scientists are now observing quantum behavior once thought to be relegated exclusively to submicroscopic realms such as entanglement, superposition of states, coherence, tunneling, and teleportation in our everyday world at a very human level.*

## Quantum Teleportation

The dual behavior of matter also gives rise to another interpretation, now termed as quantum teleportation. Teleportation can be described as the transfer of an electron—as particle, wave, or even both—from one locality to another without passing through the space in between.

William J. Chevalier (2011) described it as "the process of making a subatomic particle's physical state vanish from one place and appear in another..." The process involves dematerializing one particle at a given point in space and reconstructing it to another location, carrying with it the same information it had before the reconfiguration. C. H. Bennett, et al. (1993), in their seminal paper "Teleporting an Unknown Quantum State via Dual Classical and Einstein-Podolsky-Rosen Channels," first started demonstrating with single photons, later extending this to other material systems like atoms and ions.

Several experiments have already been performed in many laboratories across the globe proving that quantum teleportation is indeed possible. Scientists have teleported quantum bits through more than a mile long of fiber-optic wire and these experiments "have demonstrated that teleportation works in the kinds of real-life conditions that are found in telecom applications." The latest record distance reported is 143 km. or 89 miles (Stefan Lovgren, 2004).

## Unbroken Wholeness

According to this principle, particles only have their meaning and significance when taken in relation with other particles. No particles exist independently and autonomously, since on their own they cannot exist. Particles remain connected and correlated "no matter how distantly separated they may become" (Tim Maudlin, 2011).

This notion of interconnectedness was given strong philosophical and scientific foundations by David Bohm in his concept of "implicate order and unbroken wholeness." According to him, "unbroken wholeness means … you cannot analyze it. You cannot take it apart." This inseparability of quantum particles is the fundamental reality and that "relatively independently behaving parts are merely particular and contingent forms with this whole" (David Bohm & B. Hiley 1975:96, 102).

It is what "provides the framework or continuant whose presence gives unity to the whole" (K.G. Denbigh, 1975). Bohm was very much aware of the classical view that particles are isolated and act independently of each other. But, according to him, this is not so in the quantum world, declaring that this unbroken wholeness in fact constitutes the fabric of the entire Cosmos. It is this web of interconnectedness that gives wholeness to the entire cosmic system.

This notion was reechoed by Wolf (1981:178) when he explained that a reality or event can be fully understood only if taken in relation to the entire reality. Relationships, however, may be expressed in varying degrees. Neil Shubin (2013) gave the analogy of a family tree. First-degree cousins are more closely related than second-degree cousins. Thus, according to him, "Knowing the pedigree becomes the basis for understanding how different creatures are connected to one another …"

Applying this idea on the cosmic level, Werner Heisenberg (1958:107) said that: "The world thus appears as a complicated tissue of events, in which connections of different kinds alternate or overlap or combine and thereby determine the texture of the whole." There is no separation of the fundamental elements of reality. American Physicist Barbara Ann Brenan (2011:24) reechoed this view as follows:

*Quantum physics is beginning to realise that the Universe appears to be a dynamic web of interconnected and inseparable energy patterns. If the universe is indeed composed of such a web, there is logically no such thing as a part. This implies we are not separated parts of a whole but rather we are the Whole.*

Interpretations continued to flood. Fritjof Capra (1975:124) declared these notions of interconnectedness and mutual interrelation of all things and events that arise in the quantum world as the "metaphysics of quantum theory," while Gary Zukav (2001) referred to this interconnectedness as something mystical, even sublime. To quote him: "All the 'parts' of the universe are connected in an intimate and immediate way previously claimed only by mystics and other scientifically objectionable people."

Ancient beliefs have, in fact, been proclaiming this web of cosmic interconnectedness thousands of years ago. Lama Anagarika Govinda (1974:93) put it as follows:

*The Buddhist does not believe in an independent or separately existing external world, into whose dynamic forces he could insert himself. The external world and his inner world are for him only two sides of the same fabric, in which the threads of all forces and of all events, of all forms of consciousness and of their objects, are woven into an inseparable net of endless, mutually conditioned relations.*

In fact, all major world religions today have preached this teaching of unbroken wholeness through the millennia. It is a common theme in Hinduism, Buddhism, and Taoism. Sri Aurobindo (1957:993) expressed that:

*The material object becomes ... something different from what we now see, not a separate object on the background or in the environment of the rest of nature but an indivisible part and even in a subtle way an expression of the unity of all what we see.*

The concept of interconnectedness makes possible the instantaneous communication of particles, technically referred to as quantum entanglement.

## Quantum Entanglement

Quantum entanglement is closely related to the concept of unbroken wholeness. Particles are intimately connected with each other such that the act of measurement by the observer triggers an effect on other particles as a whole. According to this view, when an observer makes a measurement on a particle, an effect on the nature and behavior of another particle located in another region in space occurs, even if their distance between the particles is million of light years away.

Even barriers that lay between and among particles no longer matter. Bohm categorically declared that no signal is in fact needed at all since every particle is connected to each other as an "unbroken wholeness" (D. Bohm and Basil Hiley, 1975:94).

Moreover, the effect is not only one-way. For the receiving particle also relays its response back to the source of the information. A form of communication thus occurs between two particles forward and backward. Fred Alan Wolf (1981:74, 174) expressed this view:

> *Information travels instantaneously, backwards and forwards, in circular motion. In fact, information does not travel at all; there is no need for communication, since everything is already laid down. ... The wave spreads throughout the universe faster than light, travelling both backwards and forwards in time.*

The process is repeated in an almost endless process of communication and exchange. Time and distance no longer matter that the concept of travel may no longer be relevant. The communication exchange is so instantaneous that information, by virtue of its wave nature, travels at superluminal velocities or speeds faster than light. According to Paul Davies (1988): "it is as if the two particles engage in a conspiracy to cooperate when measurements are performed on them interdependently."

It is this instantaneous action-response relationship between and among particles that keeps the whole quantum reality alive and dynamic.

The strange quantum discoveries are also theorized to happen at the macro level. To quote de Broglie: "After long reflection in solitude and meditation, I suddenly had the idea, during the year 1923, that the discovery made by Einstein in 1905 should be generalized by extending it to all material particles..." This is mind-boggling since if this is true, we humans, if strongly connected to other humans, can also communicate with each other simultaneously regardless of time and distance.

John Luckey (2015) cited an experiment done in the United States where it was demonstrated that "light pulses can be accelerated up to 300 times their normal velocity of 186,000 miles per second." He also narrated another experiment where a group of Italian scientists accelerated the speed of microwaves at 25 percent above the speed of light.

More convincing evidences have been discovered in 2012 at the CERN laboratory confirming that subatomic particles can indeed travel faster than the speed of light. Neutrons, for example, released at the Swiss laboratory arrived at the Italian site "some 60 billionths of a second faster than if they have been travelling at the speed of light." (ibid.).

That quantum entanglement is occurring at the macro level is not quite hard to grasp. It is as if a part of one region of the line of force in the CEF that has been "plucked" (as in a guitar) resonates instantaneously to other regions across space affecting all other particles that spread out along the way.

In reality, we can also communicate at a speed faster than light, as in the case of mental telepathy. Although Fred Alan Wolf has expressed some reservations on this notion, he, nonetheless, hinted that the bizarre behavior of subatomic particles is closing in on the psychic realm. In his words (1981:201-202): "Real particles may exist, but they follow very strange orders. These orders border on what we now call psychic phenomenon."

# 3

## The Physical Realm

*In the beginning God created the heavens and the earth. The earth was without form, and void; and darkness was upon the face of the deep, and the Spirit of God moved upon the face of the waters. And God said, "Let there be light," and there was light.* - Genesis, 1:1-3

*Light is what enfolds all the universe...Light in its generalized sense (not just ordinary light) is the means by which the entire universe unfolds into itself.* - David Bohm

*In the beginning, there was Nothing,* not even a beginning*! From out of this Nothing, emerged everything we see around us today.* - Russell K. Standish

*The numerous facts collected by various sciences indicate ... that our universe had a certain beginning, from which it developed into its present state through the process of gradual evolution.* - George Gamow

*In the very beginning, there was a void, a curious form of vacuum, a nothingness containing no space, no time, no matter, no light, no sound. Yet the laws of nature were in place and this curious vacuum held potential.* - Leon Lederer

*There had been many versions explaining the creation and evolution of our Cosmos. These stories were told and re-told to us in the form of myths, legends, symbols, imagery, allegories, parables, metaphors, and theories. They were meant to help us understand who we are, what our role and mission on Planet Earth is, and what our destiny will be. There are millions and perhaps even billions of people who still believe in one or more of these stories.*

*In this chapter, I present these creation stories from the materialist perspective of the physical sciences with a sprinkling of Religion, Mysticism, and the ancient mythical beliefs of the Sumerians propagate. Today, as we know, there is today an emerging trend for Science to*

*be more religious and mystical in the same manner that Religion and Mysticism are becoming more scientific in their approach to reality.*

*I conceptualize this the long creation epic can be condensed into seven moments of manifestation paralleling the seven days of creation account in the Judeo-Christian tradition. These seven moments are, as I conceptualize it, these seven moments of creation cover the following: (1) the primal state of nothingness and oneness; (2) the appearance of the sub-atomic particles; (3) the formation of the macrocosmic realm; (4) the birth of our solar system; (5) the emergence of life; (6) the birthing of humanity; and (7) the period of rest and celebration of creation. ###*

## First Moment: The Primal State of Nothingness and Oneness

From the perspective of modern science, there existed nothing existed in the beginning. There was no time—no past or future. There was no space either, no here and there, no left or right, no up or down, no forward or backward. Matter did not exist. There were no solid particles, no physical objects, no laws to govern their interactions.

There were no forms and structures, no diversities and complexities, no dualities and opposites, no subjects and objects, no observers and observed. There was no separation, no distinction, and no differentiation. There was only oneness, then believed to be the atoms floating in the void. Atoms were the primal reality and were theorized to be indivisible and indestructible.

The term void was first introduced by the Greek philosophers Democritus and Leucippus. To them, it simply means an empty space on which atoms move around and combine with each other to form more complex substances. In the beginning, only atoms, thought to be material substances, existed. Lucretius described this by many names—'raw material,' 'generative bodies' or 'seeds' of things, 'primary particles'—because they come first and everything else is composed of them (1957:42). There was no God or Holy One to create them. Atoms simply appeared by themselves.

For more than two thousand years, this description of our beginning remained unchallenged. The Roman poet Publius Ovidius Naso, or Ovid for short, in his *Metamorphoses* written in the 8th century A.D., spoke of atoms (Marcelo Gleiser, 1997:14-15). Philosopher Gottfried Leibniz called it by another name, "monads," referring to them as the internal principle. But Ovid introduced the idea that atoms are traceable to One single source. Isaac Newton still held the same idea. Like Leibniz, he spoke of a One single source that created atoms, this time referring to it as God.

In the 1930s, the Belgian priest and physicist Georges Lemaître reintroduced this theory of "primeval atom" as a very dense state in which the particles of the whole universe existed as a huge atomic nucleus (Craig Sean McConnell, 2002:318). One of his supporters was Pope Pius XII, who endorsed his idea in a speech to the Pontifical Academy of Science in 1951.

But all this changed with the discovery of quantum physics. As it is demonstrated today, atoms are not indivisible and indestructible after all. Inside an atom are hundreds of subtler particles, the ultimate one of which is the quantum singularity identified as consisting of filaments of vibrating energy called strings. According to Davies and Gribbin (1992:107): "it is possible to picture a singularity by imagining all the matter in the universe squashed into a single point, no longer envisaging the point mass surrounded by space since space would have to be shrunk to a point as well." As physicist Dan Lincoln (2005) maintained:

> *Everything in the vastness of space was concentrated into a single point, not just a sort of point, but a quantum singularity. This had no size at all. Not only was all the matter and energy of the Universe packed into a single point, space itself was packed into the same point.*

This primordial singularity was said to be a very hot substance seething with vibrating thread of energy, strings, ready to be manifested when conditions were appropriate. According to this view, all the elements—time, space, matter, and their laws—were contained and compressed in this singularity as virtual particles awaiting for the appropriate condition to instantly manifest themselves. In the meantime, these elements remained in the state of potentialities and possibilities. But this does not mean that in the beginning there was literally nothing. For all these elements—time, space, matter, and its laws—were all there compressed into one singularity, potentially present and virtually existing awaiting for the appropriate condition to instantly manifest themselves instantly.

In contrast to the Greek philosophers' view, space was no longer empty and passive but packed with buzzing energy that concentrated on a single point, now known as the "singularity." John D. Barrow described this as the state "at which everything hits everything else: all the mass in the universe is compressed into a state of infinite density" (1994:5). It was very much alive bubbling with tremendous energy that possessed all the innate potentialities that would later create the conditions and ingredients for life to exist.

One convincing proof was Edwin Hubble's 1929 discovery that the universe is expanding. If this is so, Barrow (1994:4-5) conjectured that "when we reverse the direction of history and look into the past we should find evidence that it emerged from a smaller, denser state—a state that appears to have once had zero size" (see also Physicist Alan Guth, 1997:86).

The idea of absolute nothingness becomes inaccurate and unacceptable. According to Physicist Alan Guth (1997): "the apparently quiescent vacuum is not really empty at all, but on a subatomic level is a perpetual tempest, seething with activity." John Gribbin also described the vacuum as "a seething mass of virtual particles in its own right, even when there are no 'real' particles present."

British Cosmologist Paul Davies and Astrophysicist John Gribbin invited us to think of an empty box as a perfect vacuum—empty space (1992:142-143). But, in fact, according to them, this empty box is not really inert but "full of ghostly particles" restively appearing, interacting, and vanishing. In his book, *Theory of Nothing* (2011), Russell K. Standish applied this in terms of the computer language whose symbols only contain "0" and "1." According to him, "0" symbolizes zero information, whereas "1" signifies something, a case of "something appearing out of nothing."

Ancient religious beliefs on the beginnings of the Cosmos, likewise, portrayed the beginning as a state of nothingness, void, and darkness, symbolized by water or the deep watery substance. But it was not stagnant and dead because the Spirit of the Lord hovered over it.

A parallel account is narrated in the beliefs of the ancient Sumerians. In the beginning existed the female and male waters who gave birth to all that we now see in the Cosmos (*Enuma Elish*, Tablet I, Lines 1-10). At this state, there was no distinction since everything was united as One. The Hindus referred to the One as the Cosmos, who is at the same time the Brahman. Meanwhile, Sudhakar S. Dishit (1989) called this the Universal Consciousness, the source and origin of the existence of the entire Cosmos.

The Chinese sage Lao Tsu (Hua-Ching Ni (2003:79). referred to the One as all pervading and the Subtle Way:

> *When the subtle Way of the universe is all pervading, there is no longer any distinction between subject and object, between spiritual and material, between holy and unholy. All energies merge into harmonious Oneness.*

In summary, both Science and Religion speak of a beginning described as the primal state of nothingness and oneness packed with elements possessing the potentiality to manifest itself. This potentiality finally manifested itself in various forms in the Big Bang, a cosmic event that led to the birth of the microscopic world, the world of atoms and their sub-particles.

## Second Moment: The Birth of the Microscopic Realm

It was about 13.7 billion years ago that the primal state of nothingness exploded into a fiery explosion, an event now known as the Big Bang. In the latest version, the collision of two branes (or membranes) collided, producing a popping sound akin to the sound produced by the meeting and collision of two bubbles., likened to two soap bubbles popping out like balloons A feeble afterglow of this momentous cosmic explosion was discovered in 1965 by German-American Physicist Arno Allan Penzias and the American Physicist Robert Woodrow Wilson in the form of a steady "hiss," described as the "cosmic microwave background" (CMB) radiation, while working in their laboratory.

As the singularity erupted, it immediately gave birth to various tiny particles that swarmed frenziedly across space. The nascent Cosmos then went through a process of inflation that happened in $10^{-43}$ seconds. The intensity of the explosion inflated space in every direction at the speed of light, carrying along with it all the once virtual particles. Thus, was the beginning of our Cosmos, the beginning of time, space, and matter. At the instance of the explosion, the four forces of nature—gravity, electromagnetism, strong nuclear force, and weak nuclear force—also began to govern the motion and behavior of the newly born particles.

It took only a split of a second, too inconceivable for our mind to comprehend, for the first primary ingredients of the proverbial atoms to emerge. By $10^{-32}$ seconds, when the temperature cooled down to $10^{27}$°C, the vibrating threads of energy produced the *electrons*, *quarks*, and other particles.

Then, in $10^{-6}$ seconds, when the temperature further reduced to $10^{13}$°C, quarks coalesced to form *protons* and *neutrons*, a process that continued for the next three minutes, when heat further dropped to $10^{8}$°C. But it took more than 300,000 years for electrons to merge with protons and neutrons to finally form the primal atoms, which were initially composed mainly of hydrogen and helium. We are told by science that it was during this time that light finally glowed.

In was in this manner that these virtual particles—time, space, matter, and the various laws of nature—that were once incubated in the womb of that singularity manifested. They all became the ingredients that would give birth of the large-scale forms and structures we see above and around us today.

**Third Moment: The Appearance of the Macroscopic Realm**

The second moment produced only the lightest elements that constituted the first building blocks of matter. The third moment began around one billion years after the fiery explosion, when the temperature dropped drastically down to -200°C. But this was just the right time for hydrogen and helium gas, which abound during that time, to combine and form into giant molecular clouds, called *nebulae* (Latin word for clouds), considered to be the first large-scale structure to appear.

The law of gravity started to rule in the realm of large-scale structures, while the three other forces operated at the quantum level. The molecular and chemical combination of light elements led to the formation of various clusters of gas. These clusters formed giant nebular clouds that became stellar factories of other cosmic particles. From the molecular combinations of giant clouds evolved the proto-stars, smaller entities that gradually grew into fully developed stars. The stars in turn became the manufacturers of other elements and metals heavier than lithium.

Thus, the diverse combinations of subatomic particles in the *microscopic* realm eventually led to the construction of the *macroscopic world*, the realm of large-scale structures. In an ongoing process of inflation and expansion as well as amalgamation and coalescence, the Cosmos continued to create new stars that led to the formation of galaxies.

When the stars matured, they exploded in an event known as supernova, spewing out various atoms and molecules into space that would later form the planets, which, together with the sun and other celestial bodies, formed the galaxies. Billions of stars appeared and exploded, creating billions of galaxies and clusters of galaxies. Space became so vast that distances of galaxies are now reckoned in terms of billions of light years away from each other.

To this day, the Cosmos continues to create new stars and galaxies, expanding at the speed of light, stretching into the farthest reaches of the cosmic horizon. It took 370 million years from the time the primal star appeared, for a galaxy as big as our Milky Way to form. A galaxy may be elliptical, spiral, and irregular in shape and serves as a depository of heavy metals and elements that were thrown out of the womb of stars, as a result of supernovae. It is a massive system or cluster, consisting of stars, giant clouds, planets, and other lesser bodies, orbiting "a common center mass," all bound together by the force of gravity. Because of gravity, heavy nuclei are kept from flying out into the far reaches of intergalactic space.

Beginning in the 1990s, the Hubble Space Telescope came up with evidence that about 125 billion galaxies exist in the entire Cosmos. The majority of these galaxies are organized into a hierarchy of associations called galactic clusters, which contain many thousands of galaxies congregating within an area and are dominated by a single, giant elliptical galaxy. Groups of galaxies also congregate in an area to form bigger aggregations, known as *Superclusters*, which contain tens of thousands of galaxies. Under this grouping, the *Local Supercluster* is said to be part of a larger, more extended structure known as the *Virgo Supercluster*.

This is already inconceivably colossal, but larger scales appear when galaxies are arranged in what is called *huge networks*. One of the largest networks ever to be mapped out is the *Great Wall*, which is more than 500 million light years long and 200 million light years wide.

## Fourth Moment: The Birth of Our Solar System

The various chemical elements that exploded from the dying star continued to build large-scale structures that included eight planets rotating elliptically around the center of the Sun at varying speeds. These planets are: Mercury, Venus, Earth, Mars, Jupiter, Saturn, Uranus, and Neptune. Pluto, discovered in 1930, downgraded as a minor or dwarf planet in 2006 but reclassified again as a planet in 2017.

Gamow (1972) noted that it must have taken place within a period of 100 million years for the planets to appear and when they finally appeared, they exhibited differing mass, weight, and velocity. Neptune, Uranus, Saturn, and Jupiter are considered giant planets, while Mars, Earth, Venus, and Mars are grouped as rocky or terrestrial planets. All these planets are kept in their respective stations by the force of gravity.

The Planet Earth is believed to have been formed as a result of a massive impact collision with a giant body that eventually tore our planet into two pieces, with the bigger piece forming our present Earth and the smaller one, the Moon. The collision produced several other pieces, lesser in sizes but now formed part of the celestial bodies, namely, comets, meteors, and asteroids located in the asteroid belt between Mars and Jupiter.

Some ancient Mesopotamian texts, dating back more than 12,000 years ago, give more picturesque descriptions of the formation of our solar system. Their accounts bear close resemblance to the Sumerian creation epic that speaks of 12 gods and goddesses who were assigned their respective stations in the heavens above, and with powers to oversee and lord over them.

The description of the features and behavior of these gods and goddesses was so vivid that it led former NASA consultant Dr. Zecharriah Sitchin in his pioneering work, *The 12th Planet* (1976:214) to interpret these celestial beings as really the planetary names of our Solar System. According to him, these heavenly gods and goddesses and their planetary equivalents are:

SUN = APSU, "one who existed from the beginning;"

MERCURY = MUMMU, counselor and emissary of Apsu;

VENUS = LAHAMU, "lady of battles;"

EARTH = TIAMAT, "maiden who gave life;"

MOON = KINGU "the first-born among the gods (planets) who formed her assembly;"

MARS = LAHMU "deity of war;"

JUPITER = KISHAR "foremost of firm lands;"

SATURN = ANSHAR "foremost of the heavens;"

URANUS = ANU ""he of the heavens;"

NEPTUNE = NUDIMMUD (EA/ENKI) "artful creator;"

PLUTO–GAGA "counselor and emissary of ANSHAR;" and

MARDUK = NIBURU in the Sumerian version or PLANET X.

Interestingly enough, the creation of the planets was presented in the order of their relative distance from APSU, the SUN, the same order we know today. Sitchin noted that it is quite baffling to consider how and why thousands of years ago, the Sumerians already knew what modern science knows about our Solar System today.

Thus was the fourth moment of manifestation and with the creation of the heavens above and the earth below, the stage was now set for the creation of other living creatures—the birds in the sky, fish in the ocean, crops, plants, insects, and animals that crawl and walk over dry land.

## Fifth Moment: The Appearance Emergence of Life

Modern science tells us that life appeared because of the four elements of Nature: air, fire, earth, and water, all of which came from the primal state of energy that metamorphosed itself into atoms and molecules after the Big Bang. The common elements of earth during this time were oxygen, silicon, and water, a combination of two hydrogen atoms and one oxygen atom. The ozone later was formed by the combination of three hydrogen atoms ($O_3$). Its formation provided the conditions needed for the birth life and provided the physical layer we now call biosphere or atmosphere.

The first living cells were believed to be in the depths of the ocean. After millions of years, these simple cells became more complex giving birth to plants, fish, and other sea creatures of varied forms and sizes. As they increased in number and size, some species became courageous enough to dare living in and out of the ocean. They ventured on land and in due time began to learn how to breathe air with their gills, while others began to develop lungs.

Many became amphibious, able to live both on land and in water. But those that developed lungs learned to live and settled on land, no longer interested in going back to the sea. On land, all sorts and forms of grasses, herb-yielding seeds, and fruit-bearing trees gradually emerged after millions and millions of years.

There were fishes that learned how to fly as well as reptiles that learned to live permanently on land. In no time, crawling insects began to appear beneath the soil to fertilize the land, then flying insects like the honeybees and butterflies emerged to fertilize the flowers, crop-bearing plants, and fruit-yielding trees.

Mammoth-sized creatures, other land animals of all sizes, and the upright-walking primates also appeared, while the sky witnessed the appearance of varying sizes of birds. Some land creatures became so huge, like the dinosaurs, growing more than 30 feet long. Then, an inexplicable cosmic event occurred.

The gigantic dinosaurs that abounded the world for more than a million years simply vanished. Their sudden disappearance had been attributed to climatic change, meteoric bombardment, or an epidemic, among others. But whatever the reasons were, the dinosaurs were totally wiped out, never to reappear again, giving way to the smaller and shorter-sized mammals including us human beings to take their place and occupy the earth.

But how life emerged from the strings, leptons, quarks, or atoms is filled with mystery. Over the centuries, I found at least 11 life theories advanced by science, namely: the Evolutionary Theory; Mechanistic Theory; Vitalist Theory; Reductionist Theory; Morphogenesis Theory; Multi-Regional Metamorphosis Theory; Formative Causation Theory; Intelligent Design Theory; The Panspermia Theory; Anthropic Principle; and Biocentrism. These theories, however, are just as controversial as the ancient creation epics. None of them can explain how life really appeared on Planet Earth. Even the definition of life is fraught with controversy.

Laboratory experiments were only able to produce some conditions and necessary chemical ingredients for life to appear, but it failed to create life. The Intelligent Design Theory explains the molecular processes leading to the appearance of life in detail, but can only see a pattern that indicates an intelligent design, which is frowned upon in scientific circles because it implicates the existence of an Intelligent Designer reflects the Creationist Theory.

The same can be said of the Anthropic Principle that observes a pattern in life forms, how they are neatly designed with numerical accuracy and how even a small deviation would mean its disappearance. The Panspermia theory upholds that the essential elements of life must have been carried from other planets or galaxies riding aboard asteroids, meteors, and comets that heavily bombarded the Planet Earth over millions and billions of years ago (Rhawn Joseph, 2001:69).

The biblical account does not offer much of an explanation either. It speaks of a God who is the creator of life. Once life emerged, God created plants and crops on the third day of creation but is silent, as to how this came about (Genesis 1:11-13). It was also on the fifth day that God created the various creatures that crawl the land, fly in the sky, and the fish of the oceans (Genesis 1:20-25). But not much detailed explanation is given on how plants and creatures appeared.

The Sumerian version filled this gap. Plants known to be cultivated in Sumerian times were cereals, wheat and barley, peas, lentils, and flax. Flax provided fibers and oils, while millet, rye, and spelt were eaten. Countless orchards, fruit-bearing shrubs and trees abounded. Apples, pears, olives, figs, almonds, pistachios, and walnuts were cultivated. But this discovery only begs more questions.

How did the various edible crops, fruit-bearing trees, herbs, cereals, and animals emerge? How did the plants become animals? Did they simply evolve from the grasslands to what they are today, as Darwin's theory would have us believe? If not, who introduced them? The theory of evolution can't fully respond to these conundrums. But a much earlier version about the creation of plants and animals is narrated in what some authors believed to be part of the seventh Tablet of Creation.

George A. Barton (1937) tells of an event about how this began. Like the biblical account, the gods created living things of all kinds, the cattle and beasts of the field, as well as all creatures that move. The ancient account of "How Grain Came from Sumer" narrates that several crops were brought by Anu and Enlil from their Planet *Nibiru* and which were then introduced by Ninazu and Ninmada to the Sumerians.

When the Great Flood came during the time of Noah, the Bible tells us that the earth was inundated for seven days and it rained for forty days and forty nights, causing waters to fill the earth. All life, including the plants, crops, and animals, died. But a covenant was made between God and Noah to save him and his family that led to the construction of an ark before the Great Flood took place (Genesis, 6:18). In this ark, we are told that God commanded Noah to take along with him and his family all living things, fowls, and animals, both males and females (Genesis, 6:19-22).

But how did all those living creatures fit the ark?

Ancient theories uphold the view that it was not the living animals themselves that were taken aboard the ship, but their genes and DNA stored in sterilized tubes and kept in a sort of laboratory inside the ark.

In fact, we are told that Ea (or Enki), the great scientist of the Anunnaki who created humanity, had one of the gods accompany Noah in order to navigate the ship's landing, and to take care of the laboratory and the collected specimens.

After the Great Flood, when the soil's fertility returned, Noah, the Bible tells us, became a farmer and planted a vineyard (Genesis 9:18-21). Thus, all the living things survived and multiplied in great abundance after the flood. During the time of Solomon, various kinds of living things appeared under his care because God gave him wisdom and understanding, including knowledge about trees, animals, birds, and of creeping things and fishes as recorded in the First Book of Kings (4:22-30, 32-34).

### The Sixth Moment: The Birthing of Humanity

At the sixth moment, humanity emerges on Planet Earth. Its appearance is the one final moment in the entire cosmic ensemble. Two major accounts have prevailed through time, namely: the evolutionary theory; and the biblical as well the Sumerian accounts. Since I will be discussing this at length in the next chapter, let me just give a "bird's-eye-view" of the story of our appearance.

Physicists and biologists trace humanity's origin from the primal string and atom. All the primeval atomic and subatomic elements that swarmed across the expanding space and time were the primary ingredients. Single-celled microorganisms appeared and later evolved into plants, animals. Humanity appeared as a direct evolutionary offshoot of the maturing primates.

Anthropology and archeology trace our origin back to the Miocene Period that witnessed the presence of the *Hominoids*, gradually developing into more complex species, namely, the *Hominids*, *Homo Erectus*, then, finally, the Homo Sapiens. Based on the study of our genetic history, it was from the Homo Sapiens that the appearance of Adam and Eve has been traced.

In the biblical account of the Book of Genesis, it is hard to decipher when humanity arrived on Planet Earth. There is only a statement that God created Man as male and female on the sixth day (Genesis 1:26-

28). Meanwhile, occult literature claim that humanity appeared first as discarnate spirits before embodying in its present material form. Humanity, then, is an embodied spirit.

Contrary to scientific, religious, and occult literature, the Sumerian account contains more details as to why and how humanity was created. It speaks of the human species as the product of a series of experiments done through trial-and-error, before it was finally perfected. In particular, it narrates of a crossbreeding between animals, Caesarean births, test-tube babies, and creating a prototype that reflects the image and likeness of the creators, who were gods and goddesses.

We are also told that humanity was created to ease the burdens of the lesser gods and goddesses who toiled day and night in the underworld and underneath the oceans and in the rivers of Tigris and Euphrates. Based on the Sumerian account, the birth of humanity was quite cruel and gruesome.

It is difficult to ignore this account because it mirrors what modern science has been doing for the past few decades, e.g., genetic engineering or manipulation, stem-cell therapy, biotechnology, bioware, and many other technological advances. For all intents and purposes, this creation legend is part of the several accounts that led to the emergence of humanity on earth in the sixth moment of creation.

**Seventh Moment: The Period of Rest and Celebration**

On the seventh moment, we are told that God rested from all His works after seeing that everything He created was good. He sanctified and blessed this day. All those in the heavens—the angels, archangels, cherubim, and saints—rejoiced and paid homage to the Great Creator. In the Seventh Tablet of Creation, this day is devoted to the exaltation of the supreme deity. It is the time when the lesser gods and goddesses pay homage and reverence to the Great God of the Cosmos.

As the heavens rejoiced and worshipped God, humanity was also enjoined to do the same, a practice that continues to be commemorated to this day (Exodus, 20:8-10). We are told that the "Mighty Lord" admonished the entire humanity to exalt his name and acknowledge his powers. Mircea Eliade (1954:21) noted the admonition of an ancient Indian sage that we must also do what the gods did. He also echoed similar Sabbath (day of rest and worship) liturgical celebrations performed in the Judeo-Christian tradition, a practice observed from generation to generation (1954:23).

# 4

# Let Us Make Man In Our Image and To Our Likeness

*God said, "Let us make man in our image, to our likeness. Let them rule over the fish of the sea, over the birds of the air, over the cattle, over the wild animals, and over all creeping things that crawl along the ground." So God created man in his image; in the image of God he created him; male and female he created them. God blessed them.* - Genesis 2:26-27

*... what we are today as modern humans did not come about all at the same time, but at different times over millions of years.* - John H. Relethford

*As we look out into the universe and identify the many accidents of physics and astronomy that have worked to our benefit, it almost seems as if the universe must in some sense have known that we were coming."* - Freeman J. Dyson

*I have made it a point to treat my family to rest and recreation from Christmas through New Year. At one of our favorite beach resorts, we enjoy the tropical weather, beautiful scenery, and the delightful food and drinks that it offers. There is an oval-shaped swimming pool right in front of our cottage and I enjoy the early morning breeze as I watch my growing children go for an after breakfast swim. None of us have an excuse to be bored.*

*Aside from the beach, there is a pool table, a dart area, and a complete set of musical instruments that we can pick up anytime inspiration strikes. While strolling along the beach on the first day of our arrival, I met a friend vacationing with his family. They had arrived three days earlier and were staying in a nearby resort. A far cry from our regular watering hole, a number of bars that dot the long stretch of fine white sand, and one can hop in anytime of the day or night for a drink.*

*It was in this relaxing atmosphere that my friend and I enjoyed a ceviche of fresh raw tuna in coconut milk and lemon juice, seasoned with black pepper, finely sliced chili, onions and ginger. The plate was unusually large for an aperitif and it took a lot of space on our umbrella-covered wooden table.*

*We had a view of the best place to enjoy the stunningly beautiful, clear bluish skies above us, broken only by the white of soft clouds that changed decorated by a mixed variety of cloud formations that change their colors and shapes every now and then, providing a backdrop to the birds traveling in formation to escape winter.*

*Along the beach, there were people strolling leisurely in bright casual wear, while there were those who braved farther out into the sea—boating, snorkeling, or surfing. Barely visible and farther into the horizon were a number of fishermen scattered in tiny boats. One could literally behold a sea of humanity under a canopy of a beautifully painted celestial canvass.*

*"What an awesome Cosmos we are living in," I thought to myself, but at the same time I wondered if there are other planets whose inhabitants strolled along beaches as beautiful as ours! I thought of the many reported UFO and ET sightings. How did they appear? What or who created them? Who taught them how to defy gravity, time, and space?*

*Coline Serreau's movie "La Belle Verte" or "The Green Beautiful" (1996) and Michel Desmarquet's book on the "Thiaoouba Prophecy" (1993) fascinate me. These works have inspired me to accept the existence of an extraterrestrial world and its inhabitants in the same manner that my Christian faith has conditioned me to believe that there is life after death.*

*Moreover, I was also conditioned by science to believe that there are indeed aliens out there, but their beginnings are as perplexing to scientists as our origins. I nonetheless make it a point to say "hello" to our cosmic kindred whenever I find myself at the beach.*

*The wafting ocean breeze almost lulled me to sleep. I would have dozed off had I not been interrupted by my friend who in his usual baritone voice, remarked:*

*"The conversation we had a few months ago really made me think."*

*I felt his unexpected comments only disrupted my rumination and, honestly, I didn't want to respond to him at all, much less pursue the conversation that ended in a sour note months before.*

*Yet, I did not want to spoil his day by being uncouth when I ignore him.*

*An Asian approach when it comes to human relationships says there is wisdom in pretending to be happy and sociable. It was a lesson I learned in more than one occasion at our favorite watering hole. A common friend of ours, apparently aware that he could sometimes be quite abrasive, would loudly exclaim in between boisterous laughs:*

*"I know we are not friends, guys, but let's pretend for a while that we are."*

*We barflies found it to be effective in breaking silences and awkward moments. We have adopted the strategy among ourselves and I discovered that it is just as effective to apply within and among families as it is with friends. I once pretended to be happy in the company of my children and their friends and even sang along to their incomprehensible rap lyrics, adapting to their way of speaking and dancing.*

*Feeling young, I became one with them, to everybody's satisfaction. I even overheard my youngest exclaim how cool I was when I went to replenish my drink. Pretending to be happy gave way to genuinely being so, as I did not humiliate my children in front of their party guests.*

*Remembering the incident helped me deal with my friend. After a very long pause, I said in jest:*

*"Having sleepless nights thinking about it, I suppose?"*

*"My job," he says, "entails I travel around the world. As a European, Asia is the most outlandish place I've ever been to; I found*

*everything about it strange. When I first landed on this continent, I could not imagine myself living in this part of the world."*

*"Oh, East is East and West is West. Ne'er shall the twain meet," was my response.*

*"But, here I am now," he continues, "married to a very beautiful and loving Asian and we have a very handsome child. We often go to their province to visit their family. There, I got deeply exposed to their way of life. Today, I am more of an Asian than a European. I go home to my country every now and then, but, then, I feel I am more at home here, than elsewhere."*

*In the distance, we saw our families strolling along the beach enjoying the late afternoon breeze, headed towards us. Giving him more time to compose his thoughts, I signaled for another round of drinks.*

*But there was not enough time for the two of us to continue our reverie, since we found ourselves conversing with our families who by then were already gathered around us, inviting us to an early dinner.*
####

In this chapter, the discussion of how humanity appeared will be presented in the following order: (1) the evolutionary theory of modern science; and (2) the ancient accounts as narrated in the Book of Genesis and the Sumerian legends.

**The Evolutionary Account**

The story of our appearance can be traced far back to the Late Miocene period starting with the *Hominoids*, the species that led to the development of the *Hominines* (or Hominids in recent parlance), *Homo erectus*, and, finally, *Homo sapiens* (see Fig. 4.1). It was during this period when modern apes and humans separated in their evolutionary path, a lineage split, so to say (Bernard Wood, 2005, Roman Dunbar, 2014).

Two broad groups of *Hominoids* were discovered: the *Ramamorphs* and the *Gigantophitecus*. The two were both bipedal and tall, with the *Gigantophitecus* measuring around nine feet tall and weighing between

400 to 500 lbs. Here, we are reminded of the Book of Genesis, which tells of a time when giants roamed on earth (Genesis 6:4). *Ramamorphs*, however, were observed to be more ape-like from the waist up and only human from the waist down" (Haviland, 2000: 144, 148, 155).

With the emerging science of molecular biology in the 1960s, geneticists found that humans are 99 percent similar to the great ape species (Fred Guterl, 2013; Roman Dunbar, 2014). This is telling us that humans and chimpanzees are closely related to each other, a clear indication that we evolved from the primitive apes. Our body comes "from a common animal ancestor with the apes" Joseph McCabe (2014). These two species are considered the base or, if not, near the base, of humanity's evolutionary tree.

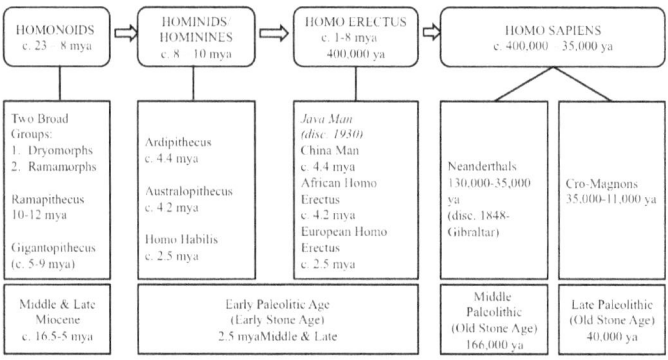

Figure 4.1. Tracing the Origins of Humanity

After the *Hominoids* came the *Hominine* group, the earliest one of which is the *Ardipithecus ramidus*, which emerged in Africa some 4.4 million years ago (mya). But this species appeared to be what anthropologists call, a "side branch in human evolution," since it did not produce any other species in the human evolutionary tree.

The genus that pursued the path of human development was the *Australopithecus*, which appeared around 4.2 mya. Though still ape-like creatures, they were observed to be "remarkably human from the waist down that had become fully adapted for moving about on the ground on its hind legs in the distinctive human manner." After *Australopithecus*, *Homo habilis*, more known as the "handy or able man," appeared about 2.5 mya followed by the *Homo Erectus*. According to anthropologists Harry Nelson and Robert Jurmain, *Homo Erectus* "should be called human" already (1979:458). A diagrammatic sketch of our ancestral lineage is given in Figure 4.2 below.

A third family that rapidly appeared in the evolutionary process around 125,000 years ago was the *Homo sapiens*, known to be the descendants of *Homo erectus*. The first to be known were the *Neanderthals*, whose fossils were discovered at various sites in Europe, Africa, Java, and China. This species was believed to be still around until 40,000 years ago. This time, scientists, using mitochondrial DNA reconstruction, found out that it was during the time of the *Neanderthals* that the common maternal ancestor ("Eve") was traced (Chris Stringer, 2012; Michael A. Cremo, 2011; Haviland, 2000:240-241). This discovery, however, needs further corroboration to appear less controversial.

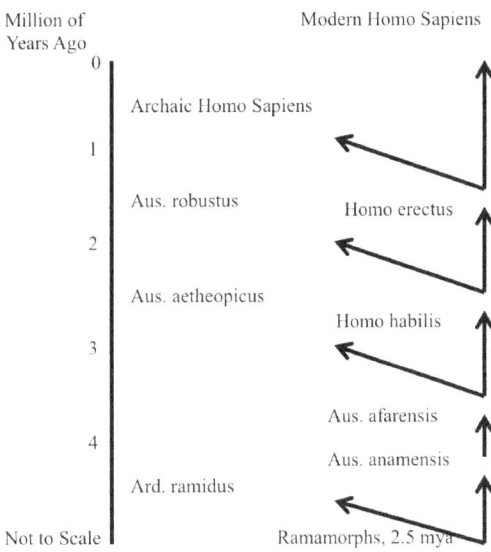

Figure 4.2. Human Ancestral Lineage

In spite of this important finding, Neanderthals were considered a side branch ensuing from the evolutionary flow of *Homo erectus*, apparently disappearing suddenly altogether between 30,000 and 40,000 years ago. This "sudden disappearance" might still change when anthropologists will be able to discover new finds between 35,000 and 40,000 years ago, a period known among anthropologists as "fossil-less years." Nevertheless, it is advanced that a genetic trace could have been left by the *Neanderthals* since there were present when the *Crog-Magnons*.

The *Cro-Magnons* emerged during the Upper Paleolithic era around 30,000 years ago. Five human remains of *Cro-Magnons* were discovered in 1868 in Southern France. Similar finds were also unearthed in other areas like Africa, Europe, North Borneo, Java, and Australia, whose remains date to 30,000 and 10,000 years ago. The Cro-Magnons were considered vastly superior to the Neanderthals— physically, intellectually, and technologically—the men standing six-foot tall on the average. But the two are distinctly different from each other physically and genetically. In any case, science's search for our ancestral lineage ends with the *Cro-Magnons*.

A far deeper issue concerning our appearance is still the subject of research to this day. Modern science still grapples with a lot of conundrums. How did the primates evolve into *Hominines*, *Homo habilis*, *Homo erectus*, and *Homo sapiens*? Did human evolution proceed in a linear fashion such that one can say that the *Homo sapiens* evolved from *Homo erectus*, and that the latter from the *Homo habilis* and so on?

The inability of scientists to answer these questions clearly and definitely has raised doubts to the theory of evolution and natural selection. Humanity cannot afford to wait for more findings to be unearthed to shed light on these concerns. At the same time, we cannot just ignore the explanations offered by our ancient ancestors.

**The Ancient Accounts**

The earliest known recorded history about the beginnings of humanity dates back to ancient Mesopotamia, around 12,000 years ago, where the first urban societies—Sumeria, Akkadia, Babylonia, and Assyria—developed. These places are located in what is today known as Iraq and part of Syria. The events that led to the creation of Man is told in the *Epic of Atrahasis*, particularly in Tablet I, which speaks about three Anunnaki gods—Anu, Enlil, and Enki—and a goddess— Mami—who were responsible for Man's birth (see also T. Jacobsen, 1987:151-166).

The *Epic of Atrahasis* came from an early Babylonian version about 1700 B.C., but is said to date back to the Sumerian creation epic, *Enuma Elish* ('When on high'). According to the account, the *Anunnaki* (directly translated as "Those who came from Heaven to Earth") arrived on Earth some 445,000 years ago, the time that the *Homo sapiens* were already roaming around earth. The Annunnaki came from another planet known as *Nibiru* (also known as "the interior of heaven") and visited our Planet in search of gold.

These extraterrestrial beings stood taller than most Earthlings and were dressed in garbs similar to our modern astronauts. They carried with them weapons that were able to inflict serious injuries and sudden death, and they displayed superior strengths, talents, and skills. It was natural for the ancient Sumerians to call them "gods and goddesses."

How the first Man was created is quite startling. Enki and his cohorts, depicted in clay tablets as dressed in aprons and holding flasks were performing experiments, perhaps in a room akin to our modern laboratory (Sitchin, 1976). As narrated in *Enuma Elish*, at par with their talents and skills, the prototype must be able to communicate, able to handle tools, understand instructions, plan, and execute tasks as good as the aliens' performances. These talents and skills were not yet manifested when the aliens came on Earth but were needed by them to relieve the gods and goddesses from the hard labor they were doing in the mines and rivers.

We are told that the Anunnaki manipulated the genes of animals and that of the *Homo erectus*, first by cross-breeding, artificial insemination, and in-vitro fertilization with other ape-like and *Homo* species. But they miserably failed. Men were born with two wings, some with four and two faces. They had one body but two heads, the one of a man, the other of a woman. They, likewise, sported several organs, both male and female. Other human figures were seen with the legs and horns of goats. Some had horses' feet; others had limbs of a horse behind, but the front was fashioned like those of men, thus resembling hippo-centaurs. Bulls bred with the heads of men; dogs with fourfold body or tails of fishes.

In the end, they realized that in order to fashion Man, divine blood or sperm coming from the *Anunnaki* was needed. And this is what they did. The procedure involved a process where the womb goddesses mixed the "essence" of the blood (sperm) of the young *Anunnaki* male with the egg of the female hominid. The fertilized egg was then implanted into the womb of a female *Anunnaki*.

This process of creation was not smooth. The period of pregnancy exceeded nine months, endangering the life of the fetus. After ten months, the womb goddesses decided to open the womb of the pregnant female *Anunnaki*. The process included reciting incantations, opening the womb, cutting the umbilical cord, among several other procedures. The operation was successful and right after the first prototype of Man appeared, a sort of mass production followed to create both males and females.

The Sumerian and biblical creation accounts bear striking resemblance to each other. Enki called the perfect prototype *Adapa*, who in the bible, is called *Adam*, and whom modern science knows as belonging to the *Homo sapiens* group. Ulligarra and Zalgarra are the Assyro-Babylonian names for the biblical Adam and Eve. It was in the city of Ashur, dating from about 800 B.C., where a record of the first two humans was found.

In the Book of Genesis, we are told Adam and Eve had two sons, Cain and Abel. But in the Book of Jubilee, an account is made that they had, in addition, other children, namely, two daughters and nine more sons (see Fig. 4.3). Since all of Noah's ancestors perished in the Great Flood, only three sons of Noah pursued the lineage of Adam and Eve. Shem is of direct concern to us since it was from his bloodline that Abram (later changed to Abraham) came forth.

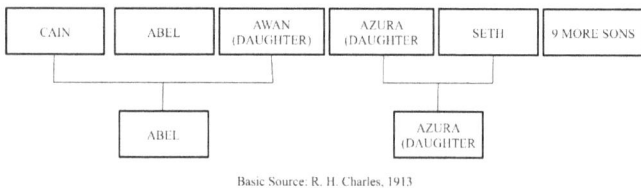

Basic Source: R. H. Charles, 1913

Figure 4.3. Children of Adam and Eve

Intermarriage within the family gave birth to Enoch and Enosh. It was from his offspring that Noah sprung forth, giving birth to three children Shem, Ham, and Japheth (see Fig. 4.4). The Biblical account did not pursue the lineage of Ham and Japheth.

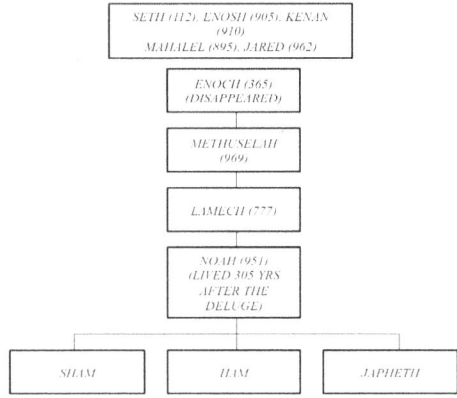

Figure 4.4. The Lineage of Seth

From the genealogy of Shem came Terah, the father of Abram, Nahor, and Haran (see Fig. 4.5). The bible also indicates that it is from the family tree of Abraham and his wife Sarah that the 12 tribes, latter transformed into nations, of Israel were formed (see Fig. 4.6).

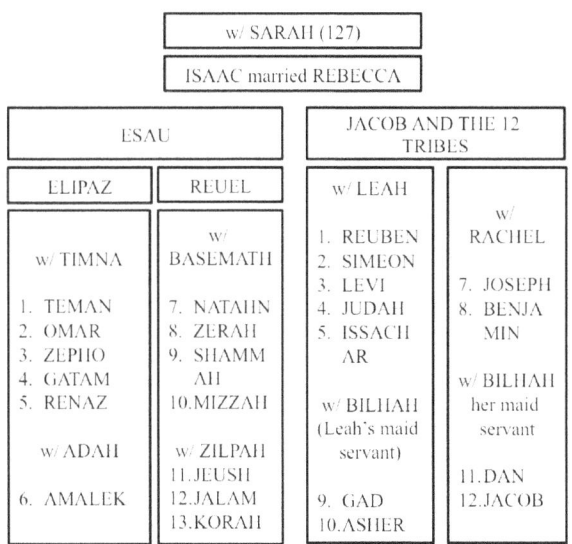

Figure 4.5. The Descendants of Shem

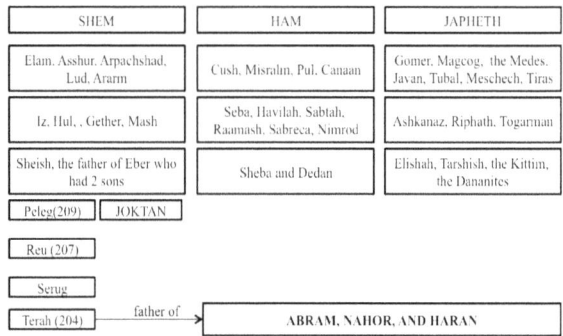

| SHEM | HAM | JAPHETH |
|---|---|---|
| Elam, Asshur, Arpachshad, Lud, Ararm | Cush, Misralm, Pul, Canaan | Gomer, Magcog, the Medes, Javan, Tubal, Meschech, Tiras |
| Iz, Hul, , Gether, Mash | Seba, Havilah, Sabtah, Raamash, Sabreca, Nimrod | Ashkanaz, Riphath, Togarman |
| Sheish, the father of Eber who had 2 sons | Sheba and Dedan | Elishah, Tarshish, the Kittim, the Dananites |

Peleg(209)    JOKTAN

Reu (207)

Serug

Terah (204) —— father of ——→ **ABRAM, NAHOR, AND HARAN**

Figure 4.6. The Twelve Tribes of Israel

# 5

## The Story of Human Civilization

*Where are we now? After going through Babylon, the Medes and the Persians, the Greeks, the Roman Empire, we are in the time of Democracies. These descendants of the Roman Empire do not function; they cannot walk.* - Russell Stendal

*Other forms of order such as those created by human beings—law, religion, the economy, society, literature, and art—are not primarily a consequence of the material order of nature ... This humanly created order has a history (unlike the natural order which is eternal) that reflects the unchanging intentional system of human consciousness—belief, desire, thought, and feeling. It is an order created by human beings and, therefore, understood by them.* - Physicist Heinz R. Pagels

*The world ... in which the presence and the work of man are felt—the mountains that he climbs, populated and cultivated regions, navigable rivers, cities, sanctuaries—all these have an extraterrestrial archetype, be it conceived as a plan, as a form, or purely and simply as a "double" existing on a higher cosmic level.... for the traditional societies, all the important acts of life were revealed ab origine by gods or heroes. Men only repeat these exemplary and paradigmatic gestures ad infinitum.* - Mircea Eliade

*Civilization comes from the Greek word "civitas," meaning city or the community in which the people reside; its inhabitants are called "civis," or what we term today as "citizens" ("civilians") to refer to those living in a particular city. In this respect, civilization is essentially a process towards "citification" or "the coming-to-be of cities." The story of our civilization, then, is a study of how we became cities, how our people organized themselves into villages, towns, and eventually cities.*

*It is a story of how, over the millennia, we developed and established our societies with all its attendant systems—economic, governance*

*or politics, social, technological, ecological, and religious—how our ancestors lived and survived as a people and how they transformed our lives into what we are today.*

*But the story of our ancient civilization is also a story of our life. It is a story of how we gathered, hunted, and produced goods, a story of how we shared and consumed these goods with others over time amidst the harshest of conditions and the cruelest of environments that, nonetheless, helped shape today's highly complex, yet organized economic system.*

*We started as food gatherers, animal hunters, and fishermen but were transformed into shepherds of flocks, livestock breeders, farmers, crop cultivators, as well as miners and metallurgists. Then, we gradually became craftsmen, carpenters, construction workers, road builders, mechanics, technicians, cloth weavers, garment workers, and dressmakers. As more and more goods and services accumulated, we were transformed into merchants, traders, financiers, as well as overseas workers.*

*A few of us became architects, entrepreneurs, bankers, industrialists, multi-nationalists, stock brokers and investors, capitalists, computer specialists, engineers, doctors and nurses, lawyers, training specialists, university professors, and corporate managers. It is also this capacity of our brain and mind that produced highly intelligent individuals the caliber of Galileo, Newton, Einstein, Spinoza, Hegel, Mozart, Beethoven, Adam Smith, and many others.*

*The story of our civilization is also a story of how we invented our work tools and equipment to access nature's abundance surrounding us. Starting with bare hands, we devised innovative technologies that included the use of bones, sticks, arrows, stones, and metals, devices that greatly improved our food gathering and hunting expeditions as well as our system of plant cultivation and animal domestication that later led to the establishment of irrigation systems, water canals, and dams.*

*The story of human civilization is a story of how we view and relate with our God or gods/goddesses, the spirits around us, and our departed ancestors as well as how we view our subhuman kindred. These religious beliefs have been recounted in several ancient literatures. At the end of the harvest, for example, we learned that Abel and Cain brought the produce of their toil as offerings to God. Abel, for his part, offered the firstborn of his flock and some fat, while Cain some choicest crops and fruits.*

*Much earlier than this, the Sumerian farmers' handbook instructed farmers to say a prayer to the goddess, or else their growing crops would be infested with insects and parasites. In the process of working for our survival and existence, organizational units evolved. Our ancestors formed hunting bands, established laws, traditions, and practices that prescribed how we ought to formulate plans and strategies, and then execute these plans in an organized fashion.*

*By doing so, they planted the foundational seeds that germinated into a cohesive political system, ensuring our survival in an otherwise violent and chaotic environment. Gradually, a primeval form of governance and type of leadership emerged.*

*The story of human civilization is also a sociological manifestation, a story of the nomads, the tree and cave dwellers, the settlers in the lowlands, and the rural folks that gradually rose to become urban dwellers living comfortably and luxuriously in posh hotels, private subdivisions, and high-rise condominiums, while the majority of the people dwell in the metropolis living in slums, pollution-ridden, and disease-stricken areas.*

*It is a story of how we relate to one other, how we organized ourselves to defend our rights as well as preserve our dignity and to live, not only in material prosperity and affluence, but more so in the field of arts, poetry, music, philosophy, and culture.*

*How we achieved this feat is quite an account to tell. To know who we are as a people, we need to know our past and learn from our successes and failures. Let me present this story from the perspective of anthropology and our ancient beliefs. ###*

## From the Perspective of Anthropology

How did civilization start? Was it a result of a long evolutionary process such that our civilization today is built from the glory and ruins of previous civilizations? Or, was civilization built by some extraterrestrial or divine beings, as ancient theorists would claim?

Like the countless missing links that marked our appearance, the development of human civilization is filled with mysteries as well. Its transformation is not continuous but marked by several gaps that span over thousands, even millions of years. Yet, inexplicably, we graduated from the foraging-hunting-fishing stage to what we are today. Joseph McCabe (2014) expressed this view as follows:

> ... 'the evolution of civilization' means the slow and gradual development of the higher and more complex institutions—the higher standards of art and knowledge and commerce and politics—which do, in spite of all their defects, raise us to a level of thought and sentiment which is as high above that of early man as his level was above that of the man-like apes.

Archaeological excavations unearthed remains indicating that human civilization began orduring the time of the *Hominids* (see Fig. 5.1). Anthropologists tell us that the first human species to forage in the forests and savannas for wild plants, fruits, and dead animal meat belong to this group, the earliest known of which were the *Ardipithecus* and *Australopithecus*. As it is today, the economic system of our ancestors was marked by a cyclical process of production, consumption, and distribution.

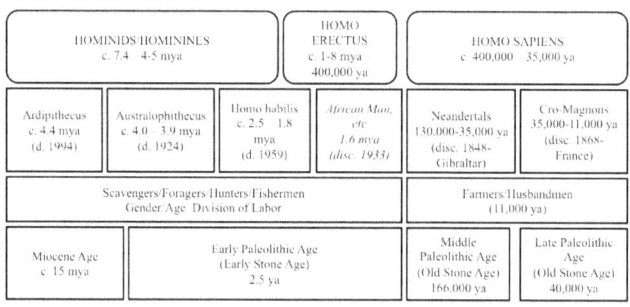

Figure 5.1. From Hunting to Farming

Our forefathers were hunters engaging themselves in expeditions, done as a family affair in places where animals congregate. It was common for them to explore the areas near their homes for days on a hunt. Hunting became inter-family when they realized it was more productive when done in groups. Meat and vegetables gathered from hunt and food scavenged became the shared property of all families, with food proportionately distributed among those who contributed in its production. There was no private property on productive resources like land because ownership was simply not in the minds of those people; there was no need of a title either.

Ownership was legitimized not by legal titles but through occupancy over long periods of time, with the occupied property being handed down from one generation to another. If the issue of ownership ever came up at all, it was held as communal, in stewardship of the chieftain. The owner-community was not an owner in the sense that they could sell the land, but were simply people who have lived in the area longer than anybody else (Nelson and Jurmain, 1982:312). This practice became a system upheld today among indigenous peoples as *ancestral domain.*

Because land was owned communally, the produce was shared with the members of the hunting bands. As the community grew larger, a system of sharing emerged. The members of the hunting band usually gave goods to their chieftain as a tribute. These goods accumulated in the hands of the chief over time, and feeling indebted to his followers, organized feasts or ceremonial occasions where the goods were redistributed to the community. Surpluses were shared with families from the other islands. In fact, neighboring communities were often invited to feasts celebrating the success of a hunt.

It became a practice that guests would return the favor by inviting their hosts in a future feast. This reciprocal practice marked the emerging economic arrangement of those times. Exchanged of goods using valued commodities like rare shells developed over time. But in many cases, barter was practiced.

Thus, our ancient ancestors' economic activities were deeply intertwined with ancient values that guided their relationships. In particular, families banded together in a system of economic activities propelled, not by the mediation of money but by the norms of cooperation, sharing, reciprocity, and interdependence.

As hunting expeditions progressed through the years, our ancient ancestors began to invent new tools and improve their food gathering and hunting skills making their hunting expeditions more effective and productive. During the time of the *Australopithecus*, people already used tools such as bones and sticks on their hunting expeditions.

This became more sophisticated with the appearance of *Homo habilis* who devised stones for slicing, scraping, chopping, cutting, and butchering meat. Nonetheless, this group of humans still got the bulk of their meat through scavenging, either waiting for animals to die or be killed by other predators.

Other major transformation that characterized *Homo habilis* was the use of a shelter, the manufacture of tools, and a marked coordination in foraging and hunting activities. There were indications already that shelters were used to store their scavenged meat as well as the foraged plants for quick processing and distribution. They also began to use fire for cooking, warmth, and even for hunting animals.

Hunting expeditions also became more organized, suggesting that a lot of planning may have been done before the actual game commenced. An ancient form of communicating and expressing their ideas through the use of guttural sounds and bodily movements may have greatly developed during this period.

The *Homo erectus* became more known as skilled hunters, no longer dependent on animal carcasses. Most famous among the fossil remains of *Homo erectus* were skilled hunters using more sophisticated and diverse stone-tool technologies for different purposes. But they continued to scavenge for food using tools made out of stone and bones. Improvements in the *tools* and *technology* of our ancestors were seen markedly with the appearance of *Archaic Homo sapiens*.

One group of *Homo sapiens* were the *Neanderthals*. They were widespread between 130,000 and 35,000 years ago and were known for their pioneering work of devising novel ways and means to hunt and kill animals for food. Archeological evidence shows they engaged in large hunting games that included bears, mammoths, and rhinoceroses through deliberate careful planning, better logistical organization, much improved hunting tools, and the use of communications during large hunting expeditions. New hunting strategies, like the use of deep ravines for cliff-fall hunting, were practiced (A. Robert, 1976; Slocum, 1975:42; D. Morris, 1967).

Then, starting more than 100,000 years ago emerged another group of *Homo sapiens*, the *Cro-Magnons*. Like the *Neanderthals*, *Cro-Magnons* still lived in caves, rock shelters, and in the open spaces and were more skilled at making hunting tools. They made spears, bows, arrows, and nets as well as fishhooks, harpoons, and needles out of stone, wood, bone, and animal fur. With their sophisticated tools, greater experience, and knowledge, our ancestors were transformed from being foragers and scavengers to farmers who domesticated animals and cultivated plants, practices that laid the foundation for agriculture around 11,000 years ago.

In effect, the Cro-Magnons manifested a transformation from nomadic way of life to one of permanent settlers (see Fig. 5.2). Beasts of burdens and animal manure were gradually introduced to increase productivity. As a result, agricultural products diversified. From the animals came the milk and skin fur for clothing and their bones were utilized as tools. Animals were used also for transportation. In no time, the production of clothing, furniture, utensils, and other materials from agricultural products created surpluses, which were traded with neighboring communities.

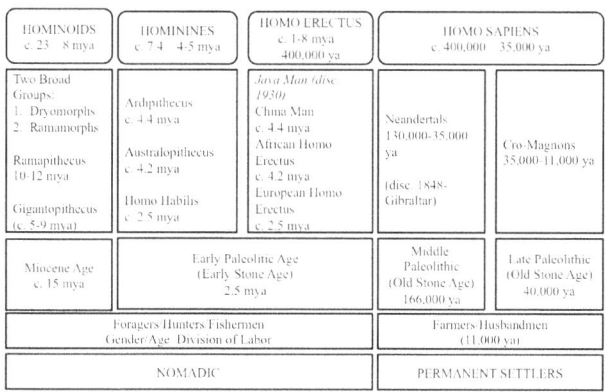

Figure 5.2. From Nomadic to Permanent Settlers

Hunting was still widely practiced, but given the difficult terrains and challenging conditions, it was done in a cooperative, rather than competitive manner. It became a communal affair paving the way to better cooperation, greater preparation and planning, coordination, logistical organization, division of labor, gender specialization, and improved communication abilities, intensifying further the solidarity of the community. As A. Rosman and P. Rubel (1998:165) noted, "cooperative endeavors in which people work communally serve to reinforce the social solidarity of the group."

A *division and specialization of labor* emerged between males and females (Lovejoy, 1981). Men who did most of the hunting were exposed to great risks and dangers from fierce and unsuspecting animals and the harsh climate. It was the women who did the foraging and gathering of edible wild plants and perhaps small mammals. They also collected sea life along the shores or fish from lakes, rivers, and streams near their homes. The tasks performed by women were less dangerous and were done in addition to protecting, caring, and nursing the younger and elder members of the family, activities which were not possible had they joined the men on hunting expeditions that often required traveling long distances over many days.

While men were away, it was practically the women's responsibility to take care of the household's daily needs, which also required planning, organization, and management skills. The food women gathered served as main staples when there were no hunting expeditions. When this supply of food dwindled, this constituted an occasion for the entire band to move to other places where other types of crops and vegetation were plentiful. The women eventually learned how to plant crops and raise animals.

Agriculture further improved. Farmers adjusted the time of their planting, growing, and harvesting as well as the breeding and raising of animals in accordance with the climate. Multiple crops and grains were grown in small gardens even as various kinds of domesticated animals were raised. With more sophisticated farming tools and implements in place, new systems of irrigation, planting, sowing, and the harvesting of seeds and crops emerged to increase land productivity.

It was also during this time that Adam and Eve engaged in a conversation with God after the two succumbed to the wiles of the snake. Zechariah Sitchin (1980) remarked that what particular language was spoken, we simply do not know, but it is surmised that, whatever it was, it could be the "Mother of All Languages." In the story of Babylon, the people spoke only one language until God dispersed humanity across several regions of the earth and made other languages to confuse them (Genesis 11:4-9).

We are not told, however, whether these latter languages really came from the primal language spoken by Adam and Eve, except that many languages today use the same words conveying the same meanings (Sitchin, ibid.).

Afterwards, people began to form different types of civilizations. Families organized themselves into communities that eventually led to the establishment of tribes, villages, and towns—starting the process of civilization (see Fig. 5.3). The formation of communities became quite pronounced with the appearance of the *Cro-Magnons*. Villages became extensions of hunting bands and farmers. While hunting was still done in faraway places, animal and plant domestication usually appeared near bodies of water, precisely because it is in these watery plains that wild plants, grasses, and seeds grew in abundance and where herds of wild cattle congregated for food and water.

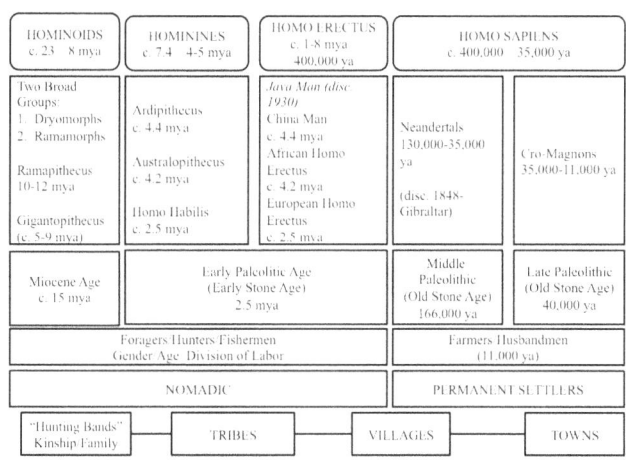

Figure 5.3. The Process of Civilization

In more arid grasslands and deserts as well as in regions where grass was abundant, families kept large herds of domestic animals and led a pastoral, semi-nomadic life. Being engaged mainly in herding, they relied on their neighbors for plant foods. Working near the rivers, it did not take long also that the permanent settlers came to discover such metals as copper, silver, and iron, transforming some of them into miners and metallurgists (see Fig. 5.4).

| Ardipithecus c. 4.4 mya (d. 1994) | Australopithecus c. 4.0–3.9 mya (d. 1924) | Homo habilis c. 2.5–1.8 mya (d. 1959) | African Man c.1.6 mya (disc. 1913) | Neandertals 130,000-35,000 ya (disc. 1848) | Cro-Magnons 35,000-11,000 ya (disc. 1868) |
|---|---|---|---|---|---|
| Scavengers Foragers Hunters Fishermen Gender/Age Division of Labor | | | | Farmers Husbandmen (11,000 ya) | |
| bipedal with dexterous hands for using tools | Designed tools for specific use | Sophistication in the construction of shelters | | Improved hunting techniques & superior technology | Elongated tools, net hunting, throwing spears, bow and arrow |
| Still spent more time in trees | Fully erect | Did their own killing Fire began to be used | | Greater planning, logistical organization More efficient social organization | Outburst of artistic expressions carvings, rock art, cave paintings, etc. |
| Spent time on the ground and savannas | Food sharing and cooperation among sexes | Developed organizational skills and planning | | | |
| Cave art and carvings, some sort of spoken language | Communicated through a combination of calls and gestures | Developed communicative skills, some sort of spoken language | | Still lacked the physical features for spoken language | Cave dwellings, rock shelters, also structures built out in the open |
| SCAVENGING, HUNTING GATHERING, FISHING | | | | PLANT CULTIVATORS ANIMAL DOMESTICATORS MINERS, METALLURGISTS | |

Figure 5.4. From Scavenging-Hunting-Fishing to
Farming and Mining

In the process of endless trading, these communities grew and prospered leading to the development of towns and cities (see Figure 5.5). It was during the Neolithic Period (c. 7,500 B.C.), the time of the *Modern Homo Sapiens*, that cities, nations, and empires exhibiting more elaborate and sophisticated forms of societal systems came about.

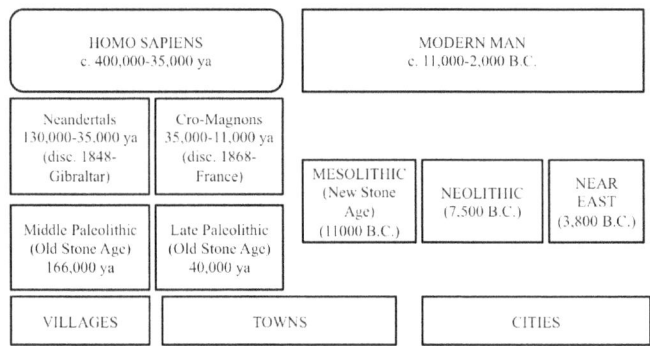

Figure 5.5. The Process of Citification

Law, education, health, and engineering became more pronounced. Laws expanded to include other areas of societal endeavors and were strictly enforced to prevent or apprehend violators. Education became necessary to hand down customs, traditions, and beliefs from one generation to another. Engineering can be seen in the gigantic structures that still baffle the minds of today's architects and engineers.

Our ancestors were also deeply religious. Anthropologists have uncovered evidence that during the time of the *Hominids*, they already espoused an intense belief in life after death and in the existence of divine beings or deities who created and ruled the Cosmos (Harry Nelson and Robert Jumain, 1982:533). English anthropologist Sir Edward Tylor (1971) cited archeological findings showing that they "had already burial rights, made amulets, and were concerned about death, evil spirits and forces beyond human comprehension."

Several findings had been unearthed confirming that our ancestors believed in invisible, transcendent beings. According to Nelson and Jurmain (1982:476), *Peking H. erectus*: "...hunted, killed, and ate his human prey elsewhere, usually bringing only the skull back to the cave ... (for) their use as ritual items or trophies. Eating the brain may not have been dietary but a magical ritual..."

*Neanderthals* were said to be the first to bury their dead as evidenced by the presence of carefully arranged tools, food, and flowers in graves. Cave and rock carvings include several objects like goat horns, skulls, and flowers in *Neanderthal* burial sites (Nelson and Jurmain, 1982:519). This discovery also manifested their creativity of symbolic thought and resourcefulness, other than holding beliefs in the supernatural (John H. Relethford, 1997:321).

In summary, our ancestors' way of life has evolved into a system that marked by interconnecting community activities that include economic, political, social, religious, technological and ecological (Fig. 5.6).

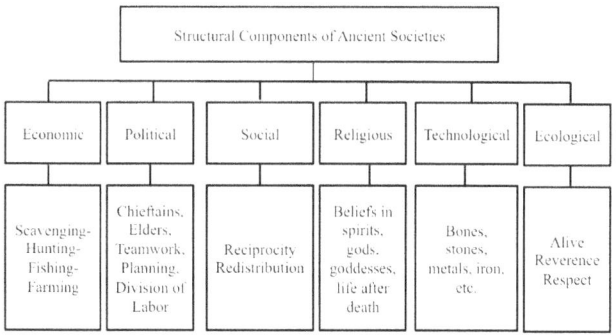

Figure 5.6. Features of Ancient Civilizations

During the earlier stages, rituals could be performed by anybody in the community. When the community grew and became more complex, religious rituals began to be performed by individuals who latter became shamans, diviners, sorcerers, witches, magicians, and priests. These religious gatherings became the center-stage of the people's community activities. Rituals and traditional banquets were frequently held in celebration of good harvests and, during times of bad harvests, where sacrificial offerings were made to appease the spirits.

Religious offerings are done in settling political differences. They became the unifying force that bound people together in peace and harmony. From *Hominines* to *Homo Erectus* and *Neanderthal*, a coherent pattern in the societal life could already be discerned, a way of life centered on religious activities around which their economic, political, social, and cultural activities revolved. Thus, our ancestors' social system eventually prepared the ground for modern civilizations to come (see Figure 5.7).

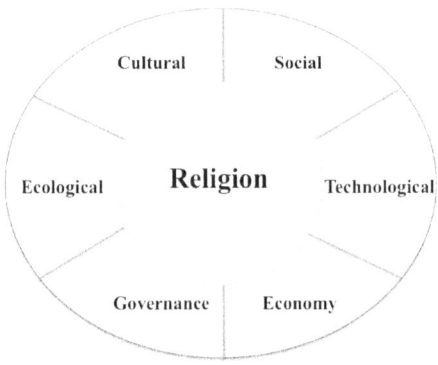

Figure 5.7. Our Ancestors' Way of Life

It was after the Great Flood, the time of Noah, that the world witnessed the emergence, rise, maturity, and disappearances of several civilizations starting from Mesopotamia. Known as the cradle of civilization, it also vanished, giving way to Sumer, which followed the same fate (see Figure 5.8).

Thereafter, several civilizations emerged almost simultaneously between 5,000 and 3,000 years in various regions across the globe like Egypt, the Indus Valley, China, and Africa. Assyria flourished between 4,000 and 2,000 years, followed by the most known ancient civilizations, namely, Greece, Rome, and Judea that began circa 3,000 years ago. All of these civilizations rose to the peak of their power and dominance only to decline and vanish.

Civilizations were found to have a life cycle of their own. Sociologists refer to this as the "sociological cycle theory." Russian Philosopher Nikolei Danilewski (1822-1985) confirmed this life-cycle duration hypothesis when he applied the theory in his studies on the various civilizations that included the Egyptian, Chinese, Persian, Roman, German, and Slav, among others.

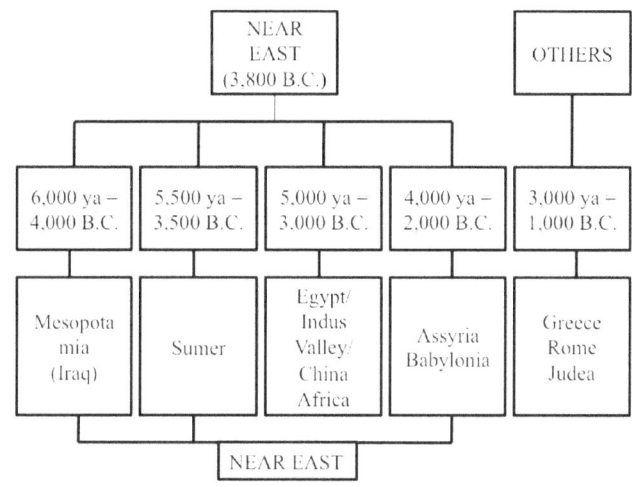

Figure 5.8. Civilizations in the Near East and Outlying Countries

## From the Perspective of Ancient Beliefs

Prior to archeology, information about our past was handed down to us in the form of sacred writings, poems, and legends. Ancient history tells us that the lesser gods known as the *Igigis*, were sent to mine gold which were then brought back to their planet *Nibiru*. The first order of business, however, was to build several landing sites for their lesser-sized spaceships to travel back and forth to the main ships stationed above the heavens. As narrated in the Akkadian "Deluge Tablets," it was the *Anunnaki's* ruling elite and aristocracy, the *Nefilims*, who did all the planning and strategizing. These landing sites gradually became the area where cities were first built.

Even before the Great Deluge, the *Nefilims* had established several cities, the most notable being those of Eridu, Bad-Tibira, Larak, Sippar, and Shuruppak. Some authors suggest that, akin to today's airports, we could just imagine, as many ancient theorists contend, that in these landing sites were command Control Centers that continually monitor and direct aircrafts. There were crews that took care of the maintenance of spaceships and managed the landing sites. It is not too wild to imagine also, as ancient theorists advance, that within the vicinity were the residential homes of the *Anunnakis*.

As a way of locating and spotting these landing sites from above, imposing structures and landmarks were built in many parts of the world to guide the spaceships entering Planet Earth. Many of these landmarks were not noticeable when viewed on land, but were seen in its entirety by the pilots maneuvering to land their aircraft. Ancient theorists claim that many of these places, especially those around the vicinity of the landing sites, may not be accessible at all to the Sumerian public in general, perhaps for safety reasons. But there could have been cities in other places where the public could go, since ancient accounts narrate of cities and civilization that began to proliferate near oceans and big rivers.

At the time before the Great Flood, the Book of Genesis tells the story of Enoch, who after living for 365 years, was taken by God to heaven (Genesis 5:23-24). The Book of Genesis does not tell us what happened to Enoch when he was taken to heaven. But details of the event are contained in *The Book of the Secrets of Enoch*, which were found in Russia and Serbia and preserved in the Slavonic language. Its translators and ancient theorists believe that Enoch's story was used by early Christians and remained important in the study of Christianity.

As the account goes, two men woke Enoch, admonishing him not to fear because they had been sent by God to accompany him to heaven (Chapter I). While with the Lord and the angels, Enoch directly received knowledge and wisdom. He was entrusted with the secrets of heaven and earth. In quick succession of events, the two men showed him the order of stars, the 200 angels who ruled them (Chapter IV), the treasure-houses of the dew and how all the flowers of the earth were made to shut and open (Chapter VI). In no time, Enoch learned about the structure of the Cosmos.

He learned science, mathematics, as well as the structure and dynamics of the constellations, stars, sun, moon, and earth (Chapters XI TO XV). In the process, he learned about the twelve gates (corresponding to the 12 months of the year) and the number of days assigned to each gate, with a total of 365 and quarter days of the solar year. He was brought to the second heaven that houses the fallen angels (Chapter VII), the place of the righteous and compassionate (Chapter IX), and other higher realms.

Then, the Archangel Michael brought him before the Lord surrounded by troops of cherubim and seraphim. The Lord summoned one of His archangels, Pravuil, the keeper of knowledge and wisdom, to bring out the books and a reed of quick-writing to be delivered and given to Enoch so the later can take down notes in his own handwriting (Chapter XXII). The angel was telling him about the beginnings and workings of the heavens, earth, and sea, their passages and goings, the thunders, the sun and moon, stars, the seasons, years, days, and hours, among many others (Chapter XXIII).

The Lord Himself, in the presence of angel Gabriel, revealed to Enoch the great secrets of God, telling him how He created everything and all the forces and that He, "the self-eternal, not made with hands, and without change" existed from the beginning even before everything was created. When everything was finished, the Lord commanded the two angels to accompany Enoch down to earth taking with him the books so that Enoch can relay what he learned from the Lord to his sons.

The Lord also instructed Enoch to distribute the books and the notes written in his own handwriting to his sons and their children and hand these down from generation to generation, nations to nations (Chapter XXXIII). Enoch, indeed, did what the Lord instructed him to do. He told everything to his sons what the Lord had commanded him to do and passed the books to them as his legacy to the coming generations (Chapter LIV).

How the people learned the secrets of the heavens, the underworld, as well as biology, chemistry, and mathematics also appears in the story of Enmeduranki, whose life parallels the life of the biblical Enoch, leading many scholars to believe that the two are the same personality.

Like Enoch, Enmeduranki was given the "Tablet of the Gods" and the secrets of the gods.

How about their knowledge of agriculture and farming? How were the plants and crops introduced?

The Sumerian account titled "How Grain Came from Sumer," narrates that Anu and Enlil brought several crops from their Planet *Nibiru*, which Ninazu and Ninmada introduced to the Sumerians. The gods and goddesses also taught the Sumerians how to farm (Black, J.A., G. Cunnigham, E. Robson, and G. Zlyyomi. 1998). The work of the Greek poet Hesiod, *Work and Days*, recorded in the 8th century B.C., is a detailed guide for farmers, helping them in their year-round activities (Samuel Kramer, 1981:65-69).

Considered as the "First Farmer's Almanac," it is narrated that a farmer, who turned out to be the god Ninurta, instructed his son about farming and agriculture. The details are quite surprising because the method and system of farming is still being practiced to this day.

It speaks of how to plow the field, what instruments to use and how to apply them, the use of the beasts of burden to help in the tilling of the land, the method and measurement when planting seeds, how to protect the growing seeds from insects and mice, when and how to reap the barley, when to winnow, and even when to synchronize farming activities with the constellations of stars. As for farm implements, it is narrated in the manuscript "Creation of the Pickax" that it was Enlil who introduced the hoe "to make room for seeds to come up."

Already during the time of Solomon various kinds of crops and fowls multiplied under his care because God gave him wisdom and understanding that exceed all the wisdom of all of Egypt which is to know everything, including knowledge about trees, animals, birds, and of creeping things and fish (First Book of Kings, 4:22-30, 32-34).

Meanwhile, how did the animals appear?

We are told in the Bible that on the fifth day of creation, God decreed to bring forth abundantly living creatures that fly above the earth, move on land and under the seas. A much earlier version is narrated in what some authors believed to be part of the seventh Tablet of Creation. Like the biblical account, the gods also created living things of all kinds, the cattle and beasts of the field, as well as all creatures that move. But when the Great Flood came during the time of Noah, the Bible tells us that the earth was inundated for seven days and rained for 40 days and 40 nights (Genesis 6:17).

But not all living things were wiped out. A covenant was made between God and Noah to save him and his family taking along with them all living things, fowls, and animals, both males and females (Genesis, 6:19-22). The Sumerian version also spoke about the knowledge of the impending Great Flood and, in view of this, Enki, the creator of Man, not wishing that all of humanity would perish, ordered Ziusudra (in the Sumerian account) or Upnapishtim (Akkadian narrative), and Atrahasis (Babyloniah version) to build an ark so that his family could be saved when the Great Flood came. He was also instructed to bring along with him living creatures, both male and female.

But who taught our ancient ancestors the extraction of metals from their ores and the modifying it for their use? Who taught the people biology, chemistry, science, and metallurgy?

The ancient manuscript "The Book of Jasher" says a certain Azazyel, from the lineage of the giants that roamed the Earth, taught the people the art and science of cross breeding. In Chapter IV:8, an account is told about the judges and rulers who taught the people "the mixture of animals of one species with the other," also referred to in Joshua (X:13) and Second Samuel (II Samuel 1:18; Chapter 8:1).

Azazyel's name is also mentioned in the Book of Leviticus (16:8) in some Bible versions. It was from the giants that people were taught the art of writing, reading, medicine, geography, meteorology, astrology, weaponry, cosmetics, as well as witchcraft, sorcery, divination, communicating with the dead, spells, incantations, snake charming, palmistry, and many other things.

With the knowledge and wisdom the gods had taught came the proliferation of human civilizations. Nations were born when Noah dispersed his sons after the Great Flood (Genesis 10:32): Modern science confirms that cities began to proliferate after the Great Deluge (see Fig. 5.9).

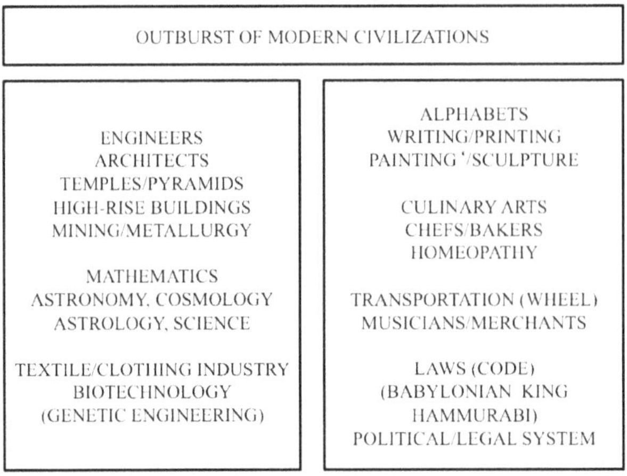

Figure 5.9. Outburst of Modern Civilizations

The scavengers were transformed into farmers, merchants, metallurgists, cosmologists, astronomers, architects, builders, craftsmen, and the like. In addition, women were no longer confined to their household chores but became working professionals engaged in several income-generating activities. In his *Legacy of Sumer*, Denise Schmandt-Besserat (1976) gave us this image of women during those times:

> *Women engaged not only in household chores like spinning, weaving, milking, or tending to the family and the home, but also were "working professionals" as doctors, midwives, nurses, governesses, teachers, beauticians, and hairdressers. . . . Women were also singer and musicians, dancers and banquet-masters.*

What do all these accomplishments mean to us? Is this the one percent difference geneticists are talking about that distinguishes us from the chimpanzees? If so, what is it in our brain and genetic make-up, Enki and the gods implanted, that catapults us into what we are today?

# 6

## The Mental Realm

*One benefit of switching humanity to a correct perception of the world is the joy of discovering the mental nature of the Universe. We have no idea what this mental nature implies, but—the great thing is—it is true...The Universe is immaterial—mental and spiritual.* - Richard Conn Henry

*...the stream of knowledge is heading towards a non-mechanical reality; the Universe begins to look more like a great thought than like a great machine. Mind no longer appears to be an accidental intruder into the realm of matter... we ought rather hail it as the creator and governor of the realm of matter.* - Sir James Jeans

*Just as there exists the outer cosmos—the physical universe—there also exists the corresponding inner cosmos of the mind.* -Subhash Kak

*The mental world and the physical world are not two separate things ... They are intertwined in a way. They're complementary; they are two different ways of looking at reality, and they are each legitimate.* - Eva Herr

*Many believe that the achievements we have attained today is primarily a product of the continuing development of our cranial abilities to know and discover something that, unlike our predecessors, enabled us to build telescopes, scanning and imaging devices, automobiles, airplanes, satellites, and computers, among many others. It is this capability of our brain that differs us from the primates. Geneticists have sequenced the human brain genome and found that our brain differs from that of the chimpanzees by only one percent.*

*Modern science maintains it is this one percent that allows us humans to perceive and discover new ideas, invent new technologies that can send us to space, and realize great achievements in arts, music, language, and literature that chimps cannot do. Our difference with the primates lies in our faculty of perception and cognition.*

*It is because of our brain and mind that empower us to produce not only symbols, concepts, and ideas but also thoughts, feelings, and dreams. It is this realm within us that brought our civilization to what and where we are today—the age of science and technology, the age of globalization and industrialization, and the age of space exploration, a far cry from those dark ages when we were mainly scavenging, hunting, and fishing.*

*To understand what it really means to be human we need to go within ourselves and examine what lies within us that empowered us to triumph and succeed far beyond the hominines we once were. We cannot fully understand how we have come to what we are today if we remain fixated on exploring and conquering the outside world. For what we have achieved over the past million years were simply expressions of our innate powers and capabilities. ###*

The human brain did not just emerge out-of-nowhere. It is a product of evolution coming from the same ingredients that produced the stars and galaxies. Over millions of years, subatomic particles combined to form molecules that in turn transformed into living cells, tissues, brain, and finally our mind (see Fig. 6.1 below).

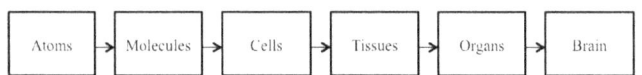

Figure 6.1. From Atoms to Brain

The first to emerge in life were single-celled organisms without cell bodies, also known as *prokaryote*. Then after another millions of years, cells with nuclei, or *eukaryotic* life, emerged from which multi-cellular life, appeared (Frank Amthor (2011). Cells in multi-cellular species began to specialize and communicate with each other, leading to the evolution of the reptilian, mammalian, and the neo-cortex.

Of the three, the reptilian brain is the oldest and first to evolve; It is responsible for all the primitive functions of our body like breathing, swallowing, reflexes, territoriality and many others (see Fig. 6.2 below).

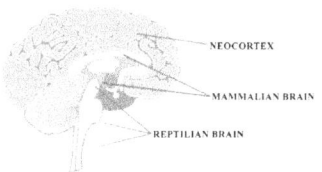

Figure 6.2. Three-Layered Human Brain: Reptilian,
Mammalian, Neocortex

The next to evolve is the mammalian brain, the part of the brain where basic emotions like love and hate, joy and suffering, happiness and anxiety, etc. are processed.

The last evolutionary stage of our brain's development is what scientists call the *neocortex* (termed "neo" because it is the latest to appear) and considered as the reasoning part of the brain and the area where art, symbols, intelligence, communications, language, memory, and free will are processed (Frank Amthor, 2011). So unique and fully developed is the human brain that only a few higher mammals and humans are known to have them.

In the late 19th century, English Anthropologist Edward B. Tylor (1971) noted that our *hominid* ancestors already exhibited this three-layered structure of our brain. Over the millennia, the size and mass of our brain underwent a series of progression (see Fig. 6.3 below). The earliest humans had brains of approximately 36 cubic inches, compared with that of the mammals with an average brain size of 12 cubic inches (Yuval Noah Harari, 2015).

The earliest beginnings of the human brain date back to the appearance of *Australopithecus*, considered as the earliest *Homo* species, whose cranial capacity (an index of brain size) measured between 310 and 485 cc., said to be 1.5 times larger than that of modern monkeys. *Homo habilis*, with a brain capacity of between 650 and 690, is the next big leap in the development of our cranial structure. Anthropologists tell us that its brain was "reorganized along more human lines" and that it was clearly "proceeding in a direction different from that of *Australopithecus*" (Haviland, 2000:178).

| Species | Cranial Capacity (in cubic centimeters) |
|---|---|
| Prototypes | 167 |
| Australopithecus | 310-485 |
| A. Afrinensis | 428-500 |
| A. Robustus | 530 |
| A. Boisie | 500-530 |
| Homo Habilis | 650-690 |
| Homo Erectus | 700-1,225 |
| Neanderthal | 1,400 |
| Cro-Magnon | 1,600 |

Figure 6.3. Development of the Cranial Capacity of the Brain

As it continued to develop, its brain structure became more and more human and less apelike. It, in fact, already started to exhibit some degree of sharing, cooperation, planning, and foresight, even as its ability to communicate began to advance through the use of some elementary gestural language (Haviland, 2000:197). The implications of this view are astonishing. Our ancestors already possessed the basic transcendental capability to choose freely from among the many opportunities given the conditions set by their external surroundings (Nelson and Jurmain, 1982:535).

The second family to appear under the genus *Homo* was the *Homo erectus*. This time, the new species exhibited a much larger brain structure (700-1,225 c.c.). Inside its brain case were signs of near-modern development, especially in the speech area. Because of this, Milford H. Wolpoff (2000) and other paleoanthropologists claim that "there is every reason to believe that *Homo erectus* was capable of human vocal language," a capability which they must have used to the full during hunting expeditions.

A third family that appeared in the evolutionary process was the *Homo sapiens*. The first to be known were the *Neanderthals*, with brain capacity of 1,400 c.c. But the improvement in the cognitive capacity of Man reached its peak in the *Cro-Magnon*, with a brain capacity of 1,600 c.c.

With these discoveries, anthropologists maintain that our ancestors in the hunting-fishing-scavenging cultures already possess the same amount of intelligence, awareness, consciousness, and free will as those in advanced industrial societies.

## How the Human Brain Operates

Today, neuroscience is now able to explain, in great detail, the size, structure, and composition of the human brain. Thanks to the rapid advance of science and technology, we are privileged to witness during our lifetime such powerful noninvasive brain imaging technologies as the magnetic resonance imaging (MRI), x-ray, computerized tomography (CT) scan, the positron emission tomography (PET), and electroencephalography (EEG). Because of these devices, neuroscientists can now scan and even produce a three-dimensional map of the various regions of the brain (Denis Le Bihan, 2014; William R. Uttal, 2011).

Neuroscientists tell us that there are three kinds of *neurons*: *sensory neurons*, *association neurons* (also, referred to as *interneurons*), and *motor neurons*, each of which performs specific functions (see Fig. 6.4 below). *Sensory neurons* react to external sensations picked up through our senses, while *motor neurons* transmit these sensations to other cell bodies that control our movements.

But it is the *association neurons* that connect all the vast neural elements (Bruce Hood, 2012). These types of neurons consists of the synapses, axons, and dendrites, which are estimated to be over 60,000 miles of nerve fibers (Sir John Eccles, 2013, D. F. Swaab and Jane Hedley-Prole, 2014).

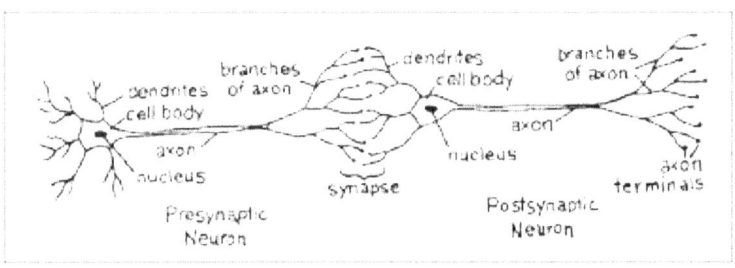

Figure 6.4. Tracing the Path of Sensory Perceptions

The primary task of neurons is to process information coming from our five physical senses (David A. Redish, 2013; Frank Amthor, 2011; Erich Harth, 1995:49). Neurons interact with each other by sending chemical and electrical signals through connecting filaments of fibers. It has been found that neurons are organized into small microscopic circuits, whose combination can form progressively larger circuits, which in turn form networks or systems (Antonio Damasio, 2010). A group of nerve cells, or neurons, assemble themselves to form several centers performing distinct and separate functions—speech, visual, auditory, motor movements, taste, touch, among others—that serve as pathways or even transmitting centers for receiving information from the outside world and relaying information from the brain to the outside world.

The processed external messages are processed by several cell bodies among the different neural networks and then sent to the corresponding sensory neural assemblies. A visual sensory neuron, for instance, only receives those stimuli that originate from the sense of sight, while the others are received by the sensory neurons, which correspond to the sense organs relaying them. When a neuron receives the corresponding sensory perceptions from other neurons, this information is automatically grouped together and combined to become one. This happens because our five sensory organs are interconnected.

Neuroscientists, like Susan Greenfield (1998:215) have shown that an idea triggered by an outside stimulus is the result of several cell assemblies and networks of neuronal circuits of varying permutations and combinations carrying different objects, representations, or ideas from the external reality, not by an individual neuron "photographing" or receiving a single image in the outside world.

All these sensory perceptions are then transmitted to the brain's central processing unit (CPU) as one message that in turn incite the formation of ideas, concepts, or symbols. When the combined activities within the brain reach its peak, neurons start emitting molecular-chemical-electrical signals that can cause surging effects on the various neural connections. Neuroscientists maintain that the patterns produced by the neural firing of electrical activities are our thoughts, ideas, and emotions. Some thoughts or emotions can be strong and intense, while others weak and faint, depending on the intensity and power of the firing. For instance, Voltaire Indigo (2015) explains that:

> *Neurotransmitters are released from synapse to synapse in order to create the emotion of happiness or sadness and these neurotransmitters are released because of the action potential in a neuron or collection of neurons.*
>
> *The action potential is the basis for human consciousness. It is a mechanical process that induces an electrical change in the brain. ...the greater the intensity of the stimulus the faster than an action potential will fire.*
>
> *The action potential is where consciousness starts.*

## The Left and Right Brain Hemispheres

Our brain is divided into two lobes, one representing the left and the other the right hemisphere. This bicameral nature of our brain performs separate and distinct functions, giving in effect at least two different interpretations of the same external stimuli. According to Robert E. Ornstein (1997:51-52), the left hemisphere is predominantly analytic, logical, and sequential in its operation and is especially verbal and mathematical. It operates linearly and processes information consecutively. On the other hand, the right hemisphere is "specialized for holistic mentation" and is largely artistic, more holistic, symbolic and relational in its mode of operation (see Fig. 6.5 below). The left hemisphere speaks, thinks, and generates hypotheses.

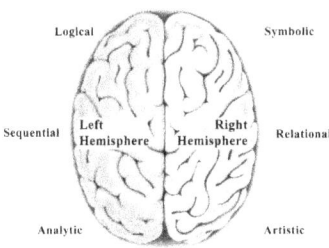

Figure 6.5. Functions of the Left and Right Brain Hemispheres

As a result, the brain can have two differing interpretations and experiences of the same external stimuli. R. Joseph (2011) illustrated this differing interpretation saying that the left brain understands the meaning of the word "outside," but it is the right brain that feels what it means to be told to step outside.

Wilson (1998:214) unequivocally argued that there's practically no one to monitor brain activities and, as Minsky opined, "there are no persons in our heads to make us do the things we want, nor even one to make us want to want" (1986:40). According to Computer Scientist John H. Holland, "there is no master neuron in the brain, for example, nor is there any master cell within a developing embryo" and thus "the *control*...tends to be *highly dispersed.*"

With thousands and millions of information assaulting the brain every second and without anyone to manage the flow of traffic, it is but natural for chaos and confusion to set in.

But how is it that we are still able to think and act coherently, sanely, and civilly?

The Cosmos must have anticipated this vexing dilemma. Neuroscientists tell us that the two brain hemispheres are physically joined together by the *corpus callosum*, a huge chain of around 200 million axons (see Fig. 6.6 below). Its main function is to make sure that the two brains work "in close cooperation and with some unity of purpose and function."

Through this connecting link, the two brain hemispheres can engage in a dialogue with each other. In fact, according to Gazzaniga, the two hemispheres are designed to interact because of the bridge of fibers connecting them (see also Physics Professor Erich Harth, 1995:123-124). This led Psychologist Robert Orenstein (1972:169) to say that we have only one brain, categorized into two lobes, each performing different functions.

Be that as it may, Julian Jaynes (1976) also noted that of all things, dialogue and communication is the last thing the two brain hemispheres do. Thus, while the opportunity is there to act together in unison, such an opportunity has not been utilized to the full.

Neuroscientists are telling us that the fibers of the two hemispheres (dendrites and synapses, axons) are not really connected to each other that the electrical signals carrying the information from one hemisphere to another have to literally jump to reach the other hemisphere of the brain. It would take a lot of conscious awareness, will, and intent on the part of one hemisphere to reach the other.

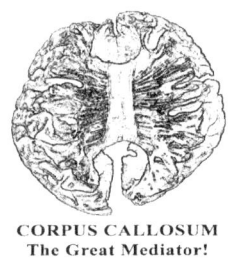

CORPUS CALLOSUM
The Great Mediator!

Figure 6.6. The Corpus Callosum

But who will start doing it? It is most unlikely that either one would initiate this since the two hemispheres are at odds with each other. Aggravating this issue is the fact that there are some other devices or networks within our brain that also do the controlling and management of our sense perceptions and cognition (Heinz R. Pagels (1990:225). Francis Crick had only to remark that this is one area that future studies of the human brain will have to address.

## The Human Mind

The mind is invisible, inaudible, massless, and weightless (G. Ryle (1949). One cannot locate it in any part of the body. But brain scientists advance that the mind exists. It is the product of what the brain does, an emergent phenomenon. It is a world of its own, beyond the reach of gravity, time, and space.

The mind, while emerging from the operations of the brain, represents the nonphysical dimension of our reality. It is nonphysical because it does not possess the physical attributes of solidity, position, and location. It manifests itself in the form awareness, creativity, and feelings.

But how is the mind able to process ideas, images, thoughts, and emotions? How do the physical elements of atoms and molecules able to produce nonphysical states?

Neuroscientist Antonio Damasio (2010) explained that the various neural circuits of the brain form and organize themselves into huge networks. In his words:

> *Minds emerge when the activity of small circuits is organized across large networks so as to compose momentary patterns. The patterns represent things and events located outside the brain, either in the body or in the external world...*

When the various processing units of the brain receive neural impulses, they ignite a reverberating effect that can activate the nerve cells in the various regions of the entire brain. It is through this process of neural firing that one region of the brain may be able to recall and construct thought patterns and emotional states from stored data or experiences, processing it alongside external stimuli (Bruce Hood, 2012).

Neuroscientists and psychologists concur that, like the brain, the mind is also bicameral in nature, consisting of the conscious and the subconscious/unconscious mind (see Fig. 6.7). We are told that each of these parts is designed to perform specific functions. The conscious mind mirrors the left hemisphere of the brain, which is rational, reductionist, logical, critical, orderly, and systematic, while the unconscious or subconscious mind represents the right brain, which is creative, artistic, symbolic, nonverbal, holistic, and emotional.

Populating the unconscious mind are our long-adhered beliefs and past experiences, which are said to be products of the external conditionings we received from our elders, parents, peers, schools, churches, media, and other external agents.

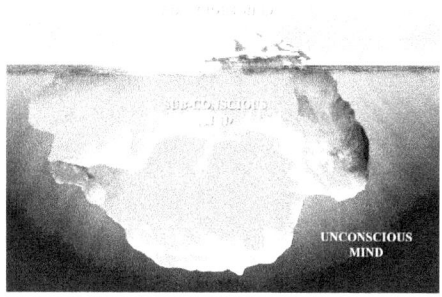

Figure 6.7. Conscious and Subconscious/Unconscious Mind

The conscious part is said to influence only around 10 percent of our thoughts, actions, and behaviors, while the subconscious/unconscious mind controls the remaining 90 percent. It is for this reason that our mind is often likened to a floating iceberg. Its visible peaks correspond to the conscious part, while the portion that lays hidden, submerged underwater is our subconscious or unconscious. But the relationship between the brain and the mind used to be a tricky issue.

The view of Philosopher Rene Descartes prevailed for quite some time. He considered the body (brain) as a complex automaton (*res extensa*), a machine, and the mind (*res cogitans*) an independent nonmaterial entity with rules of its own. In his view, the nonmaterial mind cannot interact with the material brain. But the dominant view today is that there is a close correspondence between the physical brain and the mental world (Math Professor William Kingdom Clifford, 1845-1879).

W. Uttal (2011) underscored the observation that when either one of them is not functioning or is defective, the operations in the mental world is disrupted. He wrote: "Everyone knows that the brain is in the head and that the mind is the product of the brain—no brain, no mind; malfunctioning brain, defective mind."

While the mind represents the mental world, it is also subject to the laws of physics. It obeys the laws that govern our atomic and chemical elements. But evolutionists are unable to explain how atoms and molecules are able to produce cells, tissues, and the various physical organs like the brain and know which part of our body they will finally go to and settle down. No clear explanations are given on how the various organs in our body act in unison to keep us intact and sane.

There is a deeper paradox lurking behind this consciousness-physical-mental conundrum. If the physical laws govern the behavior of everything in the Cosmos, including us, then, our behavior should be completely determined. Our mission, purpose, and future have already been determined. But how do we explain our experience of freedom? We know in our daily experience that we always make choices. We are free to make choices of what food to eat from among the almost limitless alternatives available to us.

We know that our choices could be constrained by our ability to acquire a certain brand of food as well as other contributory factors like taste, age, culture, religion, or gender. But given all these constraints, we still do the choosing. Is freedom real or just an illusion, a product of our mind's imagination? What do the physical sciences say about freedom?

**Free Will and Determinism**

There are many decision-makers inside us: the atoms and molecules in our body that are forced to obey the laws of physics; our genes that condition us to our past beliefs and experiences; our left and right brain hemispheres that are at odds with each other; and our subconscious/ unconscious mind that decides for us without our prior knowledge and approval. David A. Redish (2013) expressed the unsettling view that, "...humans (like other mammals) are mixtures of many decision-making systems, not all of which agree with each other."

From the very beginning we were already predetermined, forced by the physical rules of engagement.

We entered into the Cosmos with our role, mission, and future already set at the time of the Big Bang. Our experience of freedom is simply an illusion; in real life, what and how we think, feel, and do were scripted a long time ago. We are only to play our role and execute the mission that the Cosmos has set for us.

Neuroscientists had performed two experiments giving compelling evidences that we are not free to choose. In 1976 in Germany by Kornhuber and his associates subjected several individuals to a controlled environment and the results indicated that "doing" precedes the individual's awareness of what he/she is doing by over a second (R. Penrose, 1989:439-444). The second was a series of experiments performed in 1979 by Neuroscientist Benjamin Libet and John Eccles.

In these experiments, they monitored brain and muscle signals of individuals who were instructed to tap a finger whenever they already decide to act. The results also clearly showed that the brain executed an act several hundred milliseconds before the individual made the decision to act.

Neuroscientist Michael Gazzaniga had also been performing brain studies and experiments for the past 30 years. The results of his studies re-echoed the findings reached by Libet and his colleagues, i.e., we, as conscious human beings, feel we are the ones making the choice first when in reality it is the brain that does all the work for us. In his words (1998:63):

> By the time we think we know something—it is part of our conscious experience—the brain has already done its work. It is old news to the brain, but fresh to 'us.' Systems built into the brain do their work automatically and largely outside of our conscious awareness. The brain finishes the work half a second before the information it processes reaches our consciousness.

Since then, the same experiments have been repeated several times in many laboratories yielding the same results: we are completely determined beings and our future has already been fixed.

Herein lies the mystery of the free will-determinism debate. It is hard for us to accept that we are predetermined. In reality, we have the choice on what action to take; we are free to decide what we want to think; we are free to construct the kind of life and future we want. We cannot simply deny our experience of free choice.

With the emergence of quantum mechanics, modern science has been successful in introducing the concept of free will into the dynamics of physics through the conscious will and attention of the observer. Quantum Physicist Henry P. Stapp explained that the observer's conscious attention was determined not by the brain and that there is nothing in quantum mechanics that determines what the attention is going to be (https://youtube.com/watch?v=ZYPjXz1MVv0).

According to him, it is not the brain that brings about attention, but it is attention that affects what is going on inside the brain. Speaking on emotions, for instance, Voltaire Indigo (2015) explains that:

*The chemicals for our emotions were already decided for us before we became aware of our emotion but what we can control is how much attention we pay to our emotions and hence we feel that we can pick and choose how to feel if you choose to focus on something else that's more positive like a past memory.*

Shan Gao (2014) pursued this role of consciousness in the context of the relationships between the observed object, the measuring device, and the observer. According to him, there are two aspects involved: "(1) the physical interaction between the observed object and the measuring device; (2) the psychophysical interaction between the measuring device and the observer." He described the relationship as follows:

*In some special situations, measurement may be the direct interaction between the observed object and the observer. Even though what physics commonly studies are the insensible objects, the consciousness of the observer must take part in the last phase of measurement. The observer is introspectively aware of his perception of the measurement results.*

Our conscious awareness and willful action to perform the experiment or not reflects our free will and choice. It is this conscious will and focused attention of the observer that is responsible for the emergence or even transformation of reality. In a similar argument, Fred Alan Wolf explained that our choice and action of the kind of reality we want to manifest is pure and simple an expression of our free will. In his own words (1981:141):

*The experiences we call "reality" depend upon how we go about making those choices. Every act we perform is a choice, even if we are unaware that we have made a choice. Our unawareness of choice at the level of electrons and atoms gives us the illusion of a mechanical reality. In this way, we appear to be mere victims subject to the whims of a 'higher being.' We appear as victims ruled by a destiny we did not determine.*

In this manner, Eugene Wigner (1967) advanced that: "The human body can also deviate from the laws of physics. We can influence the formation and pattern of atoms and molecules." Neuroscientist Michael S. Gazzaniga (2011) reinforced this view:

*I will maintain that the mind, which is somehow generated by the physical processes of the brain, constrains the brain. Just as political norms of governance emerge from the individuals who give rise to them and ultimately control them, the emergent mind constraints our brains.*

In this respect, the future is not uncertain anymore. In fact, Physicist Pagels (1990) opined that it is our anticipation of the future that inspires us to make choices. In his words:

*Our sense of free will is intimately associated with our ability to perceive the future course of our actions. Unlike animals, ...we can imagine the distant future, and we can make choices that affect the future.*

It is this element that confers on us human beings the "element of inventiveness" or that "element of creativity, an ability to bring forth that which is genuinely new, something not already implicit in earlier states of the Universe" (Davies and Gribbin, 1992:42).

With the introduction of the human observer, a subjective element comes into the picture of reality. Consciousness, will, and attention are expressions of the observer's freedom of choice. Philosopher of Science John R. Searle (1984; 2004:88, 92) said that freedom, then, can be compatible with physical determinism and, according to him, "To say that they are free is not to deny that they are determined; it is just to say that they are not constrained. We are not forced to do them." This idea of compatibility or complementarity between freedom and determinism appears in the dialogue between Quantum Physicists Werner Heisenberg (WH) and Neils Bohr as follows (W. Heisenberg, 1958:91).

WH: *"You keep referring to the individual's free choice and you compare it with the freedom with which the atomic physicist can arrange his experiments in this way or that. Now the classical physicist had no such freedom. Does that mean that the special features of modern physics have a more direct bearing on the problem of the freedom of the will? As you know, the fact that atomic processes cannot be fully determined is often used as an argument in favor of free will and divine intervention."*

BOHR: *If we speak of free will, we refer to a situation in which we have to make decisions. This situation and the one in which we analyze the motives of our actions or even the one in which we study physiological processes, for instance the electrochemical processes in our brain, are mutually exclusive. In other words, they are complementary, so that the question whether natural laws determine events completely or only statistically has no direct bearing on the question of free will.*

Brain Scientist Edward O. Wilson posited a rather novel idea when he advanced a definition of free will as "freedom from the constraints imposed by the physiochemical states of one's own body and mind" (1998:232). If we subscribe to this view, as I do, freedom thus means a continuing freeing away from the constraints of the lower levels of existence. As the exercise of freedom increases, the higher one ascends to another level and, conversely, the less one becomes determined by the lower levels.

# 7

## The Psychic Realm

*... the mind of man is not only able to be aware of cosmic knowledge, but also that in some mysterious way there seems to be a cosmic initiative seeking to make itself known to man. This calls for the mind that is not only seeking but also willing to be sought.* - Edgar N. Jackson

*Within the last fifty years the human mind has been awakening slowly to the fact that there is a world, invisible to ordinary powers of vision, existing in close juxtaposition to the world cognized by our material senses.* - Nizida

*There are mysteries at the outer boundaries of our science, matters that we cannot hope to explain in terms of what we already know. When we explain everything we observe, it is in terms of scientific principles that are themselves explained in terms of deeper principles. Following this chain of explanations, we are led at last to laws of nature that cannot be explained within the boundaries of contemporary science.* - Steven Weinberg

*Because we are so connected with the living fiber of the universe, we are also extending our roots into even stranger places ... into the impressions and feelings of people who are near to us emotionally or psychically, and even unto various 'non-human sources of nourishment.'* - Percy Seymour

*Our psyche is set up in accord with the structure of the universe, and what happens in the macrocosm likewise happens in the infinitesimal and most subjective reaches of the psyche.* – Carl Jung

*Whatever you do at any moment of your Life, you cannot, for a single instant, lose your psychic ... connection with Life itself, which will go on at any subsequent moment, for all ages and for all times.* - Oris Oris

*My series of immersions into the paranormal world did not only convince me about the existence of the psychic realm. I realized that events occurring in this realm influence our search for meaning and purpose in life. I grew up in a community that until now believes in the existence of spirits, ghosts, fairies, and elves.*

*These discarnate entities are everywhere around us. We protect them and they in turn protect us. We've learned to live and co-exist harmoniously, respecting our respective spaces and sharing the same abundance of Nature.*

*I had the opportunity to attend and participate in séances performed by mediums, psychics, mystics, and those known to be instruments of some divine entities. Highly sensitive mediums and mystics communicate with discarnate spirits either for purposes of transmitting or receiving messages about life and the future. Quantum science says this is possible because everything is energy, expressed in various forms and levels of manifestations. What mediums usually do during séances is to synchronously fine-tune their frequency to that of the discarnate spirits. Once successful, messages are usually transmitted in the form of automatic writing, Tarot card reading, lifting of tables, moving of objects, remote viewing, or speaking directly through the medium, or sometimes speaking in tongues, also known as glossolalia. This is the reason why, until today, those who possess psychic abilities were either regarded as super-humans or messengers of God.*

*In ancient times, psychics were known as Oracles (and we have the famous Egyptian tale of the Oracle of Delphi) who formed part of the inner chamber of Kings and Emperors, and were often consulted first before any decision and action were made. Today, we know of Presidents, world leaders, that regularly consult with psychics for guidance.*

*Is this realm for real? Or, is this merely a product of one's wild imaginations, a world of make-believe?*

*Yes, it was hard to swallow at first but it gave me an inspiration to do more serious research about the subject. As a result, my experience, coupled with an interdisciplinary approach to understanding the unknown, erased whatever doubts and fears I once had about the world of the paranormal. The psychic realm is for real; it forms an essential property of the Cosmos.* ###

## Classical Science

Conventional science has its natural aloofness with the paranormal because it is beyond our linear and causal experience of time, space, and matter (Robert E, Ornstein, 1997:92; Madhavi Ghare, 2007). Harvard graduate Gary Zukav (2009:331) expressed an observation that "Psychic phenomena have been held in disdain by physicists since the days of Newton." "In fact," according to him, "most physicists do not even believe that they exist" (see also Love Scott, 2014).

The classicalists maintain that only our physical senses can be relied upon to know and understand reality and there are no such things as extrasensory perception and cognition. If the psychic realm exists at all, it is believed that its origin is simply traceable to the small particles that produced the atoms.

But unknown to the public, many classical thinkers have their hidden, even dark, history. The most celebrated case perhaps goes to the couple Marie Currie and Pierre Curie, both recipients of a Nobel Prize in discovering radium. Trying desperately to discover the source of energy that would reveal the secret of radioactivity, the two deeply immersed themselves in paranormal activities by attending séances with the medium Eusapia Paladino (Michael Cremo, 2011). It is in these sessions that the couple experienced and witnessed numerous paranormal events, leading them to conclude that these phenomena are for real and that the psychic realm really exists.

For Sir Alfred Russel Wallace, who cofounded Darwin's theory of natural selection, there is something more than meets our physical senses; there exists a spirit world. But the psychology and environment during their time was not ready for this revolutionary idea. Nonetheless, he teamed up with Sir William Crookes, a Nobel laureate in physics, and shifted his attention to the paranormal world.

In their experiments, they concluded that humans are not the only inhabitants of the Cosmos; there are spirit beings out there that are in continuous contact with us. Cremo quoted Wallace's findings, in his book *Contributions to a Theory of Natural Selection*, that:

> *... a superior intelligence has guided the development of man in a definite direction, and for a special purpose, just as man guides the development of many animal and vegetable forms.*

Attending séances conducted by mediums who communicate with discarnate spirits, Wallace narrates that they witnessed several paranormal phenomena like miraculous healings, apparitions, materializations of physical objects, levitations, clairvoyance, elevation of tables, movement of objects, and other such strange phenomena (ibid.).

Nowadays, in spite of its alleged weirdness, parapsychologists who have done serious laboratory studies on psi phenomena concur with the psychics that the paranormal world is as real as the physical and mental worlds (Arthur Ellison, 2012). Laboratory-based researches performed since the 1930s proved that telepathy, clairvoyance, precognition, and psycho-kinesis or the ability to move objects with the power of the mind are indeed real; the psychic world is one cosmic fact (Camille Flammarion, 1907).

Voltaire Indigo (2015) speaks about our five physical senses: the visual system (sight); the auditory system (hearing); the somatosensory system (touch); the gustatory and olfactory system (taste, smell). Parapsychologists are also talking of other senses parallel to our physical senses, now more commonly known as ESP. These paranormal senses are labeled in French as the five clairs: clairvoyance, clairaudience, clairsentience, clairscent, and clairtangency.

But the aloofness towards paranormal events is only true to the Western mind, since to the Easterners, these paranormal events are considered normal. Nevertheless, the Westerner's belief and acceptance of the psychic realm has been steadily growing over the years.

## Quantum Physics and Its Psi Implications

Quantum physics has encouraging discoveries. In the quantum world, strings are mysteriously entangled with one another that any disturbance from one space-time continuum influences the other realms instantaneously, without any regard to location and distance.

This baffling event has been termed variously as quantum jump, teleportation, tunneling, bilocation, omnipresence, interconnectivity, and oneness, as if intentionally designed to mimic psychic occurrences. Quantum physics argues that what is happening in the microscopic world can happen at the macroscopic level. We humans can also potentially tunnel through a wall, bi-locate, teleport, or be omnipresent because at our most fundamental state, we are made of molecules, atoms, and subatomic particles.

Modern science argues that we can only detect between 20,000 and 30,000 cycles per second of vibrations; outside this range are the ultra-high and ultra-low sound frequencies, which we cannot hear without the help of scientific gadgets like a transistor radio or television (see Fig. 7.1 below).

In the case of sight, the light that is visible to us consists of electromagnetic energy that covers a broad spectrum from the infrared to the ultraviolet rays. But our eyes can only detect a miniscule part of the entire electromagnetic spectrum and this is the part we see as colors in a rainbow.

Figure 7.1 Various Frequency Levels

Outside this narrow range are many other forms of light and energy that we cannot see, yet we know that they exist because modern science has invented devices for us to detect them and even enjoy the benefits they bring in our daily life, e.g., the use of microwaves, cell phones, televisions, transistor radios, and so on (see Fig. 7.2 below).

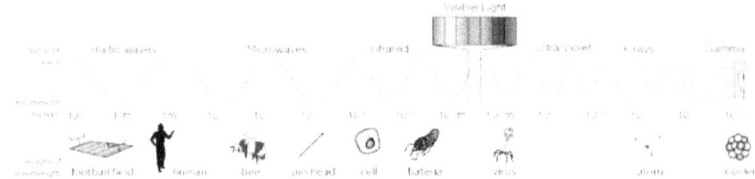

Figure 7.2. Varying Light Spectrums

This is the reason why mediums prepare themselves to reach that state of altered consciousness by synchronously attuning their frequencies to the psychic realm (see Figure 7.3). At zero Hz level, the brain is pronounced clinically dead and is registered as a flat line in the *electroencephalogram* (EEG) printout.

The *beta level* represents the conscious mind where an individual is fully awake and may be aware of what is happening within and around him. Below the beta level is the *alpha level* where the mind is in a state of relaxation and calmness; it is this state where there is more or less an absence of thought and emotion.

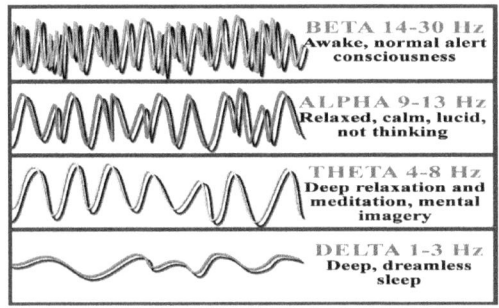

Figure 7.3. Beta, Alpha, Theta, Delta

At the *theta level*, the individual is said to attain an altered state of consciousness, achieved during sleep, dreaming, ecstasy, and other mystical experiences. Yogis and psychics attain this state of consciousness during an extended period of deep meditation. While fully alert of what is going on, it is in this mental state that paranormal phenomena are experienced to happen. This is the reason why psychics and mediums also prepare themselves by going into deep meditation and ecstatic state before they conduct séances.

Indeed, our knowledge of the Cosmos is still very much incomplete and limited. But as Love Scott (2014) explained: "We learn ... from physical science, that we are living in the midst of a world invisible to us." There are other dimensions out there and while they can't be perceived by our ordinary physical senses, this doesn't mean that they don't exist. Scott also expressed that there could be inhabitants in these realms that could have the ability to communicate with us through our five physical senses. In his words:

> ...it is not impossible that there may be living upon the earth a class of beings, also invisible to us, endowed with a wholly different kind of senses, so that there is no way by which they can make themselves known to us, unless they can manifest themselves in acts and ways that can come within the range of our own order of sensations.

Unlike the physical and mental realms, the psychic realm has its own frequency or vibrational level that can be accessed through our extrasensory perceptions (ESP). The most aggressive thinkers that study the psychic realm are the parapsychologists who are quite meticulous in doing experimental studies in controlled environments using the standards of scientific theories and the rigor of empirical methods. The findings of their studies strongly indicate that psi abilities and events, having passed strict scientific investigations, can no longer be labeled as spurious and fancies; in fact, parapsychologists believe that modern science, if it is really true to its image of being open and accepting on those that passed the rigor of empirical testing and experimental replications, ought to regard these psi research findings as authentically scientific facts.

Even then classical physicists are still reluctant to include in their research projects the subject of paranormal events. Radin (2010; 2016) deplores the fact the physicists even wait for their retirement before they are able to really expound on the significance of quantum physics in explaining the nonphysical or psychical world without fear of reprisal (Radin, 2010, 2006).

Historian and Philosopher of Science Neil Grossman from Indiana University, writing in the Foreword of Philosopher and Economist Chris Carter's work titled *Science and the Near-Death Experience: How*

*Consciousness Survives Death*, echoed the attacks of some prominent thinkers against those who still have no knowledge of these mounting data on paranormal occurrences as intellectually irresponsible. In his words (2010):

> *Given that there is a large body of empirical data that (1) is highly relevant to this question and (2) has convinced virtually everyone that has taken the time to examine it that materialism cannot explain it, I find myself agreeing with Kelly, Greyson, and Grosso that it is intellectually irresponsible for a philosopher or psychologist to be ignorant of this data.*

Grossman is of the opinion that this ignorance is more due to the tendency of scientists to avoid information that belittles their long-adhered beliefs; it is also in this manner, he says, that materialists "steadfastly ignore and ridicule the data from parapsychology and survival research" (*ibid.*).

Everything, however, is not hopeless. For the many known discoveries and findings in the world of atoms might be able to open a portal that could give way to a serious scientific investigation about the existence of a world that lie beyond our four-dimensional realm of space-time. Indeed, this came with the discovery of the *M theory*.

The *M theory* in quantum physics gave birth to the parallel universe theory that attests the existence of several dimensions out there. In fact, physicists believe that there could be an infinite number of universes, parallel to ours interacting with our universe in a very subtle way (Alex Vilenkin, 2006; Anna LeMind, 2016). Interestingly, this new theory, also known as "Many Interacting Worlds" (MIW) theory, offers new interpretations of paranormal events happening in the quantum world that are also closely parallel and similar to those of Religion and Mysticism (Brian Nelson, 2014).

So far, science has discovered 11 extra space dimensions that are believed to lie parallel to each other. These parallel universes look like bubbles that are continually in motion, moving in ripples in varying patterns and frequencies, like giant turbulent waves in an ocean. Physicist Bert Ovrut maintains that they move away from or bump into each other. They could be right next to us, but we cannot see them. Ovrut says it's like reaching out to them, but unable to touch and grasp anything because we are simply trapped in our four-dimensional universe.

The *holographic view* of the universe that emerged as an offshoot of the *M theory* also sheds light to the anomalies happening at the quantum level that include teleportation, mental telepathy, bilocation, tunnelling, instantaneous communication of distant particles, and the dual nature of subatomic particles, among others. In this respect, it is able to explain the occurrence of paranormal events and phenomena in the psychic world. As Michael Talbot (2012, 2007) explains:

> *Numerous researchers, including Bohm and Pribram, noted that many para-psychological phenomena become much more understandable in terms of the holographic paradigm... The holographic model allows us to conceptualize phenomena that have remained on the fringes of science -- synchronicities, psychic experiences, UFOs, poltergeists, spiritual experiences, states of higher consciousness.*

## Religion, Mysticism, and the Paranormal

Some religions, Christianity included, view the psychic realm with contempt and consider paranormal activities as witchcraft and works of the devil. Its practitioners are even burned at stake alive in full view of the public. But if one comes to think of it, religious believers have really nothing to be hostile against the paranormal realm. Targ and Jean Houston (2010) asserted that prayer, which is a form of communication exchange between the individual and higher entities, had produced so many miraculous results that defy the materialist view of reality. Several documented studies had emerged in recent years that point to the efficacy of prayer, particularly in the area of healing.

We also know of many Christian mystics and saints performing the gifts of prophesy, miracles, discerning of spirits, and speaking in tongues. St. Theresa of the Child Jesus was said to had been always in her ecstatic state and altered state of consciousness. Padre Pio, known for his *stigmata*, was oftentimes seen in more than one location simultaneously, e.g. celebrating the Holy Eucharist while at the same time ministering a sick person thousands of miles away.

St. Francis of Assisi was known for his ability to communicate with the birds. In the 17th century, St. Joseph of Cupertino was known as the patron of aviators for his frequent ecstasies and levitations especially during the Holy Mass celebration (Deepak Chopra, in Radin, 2013). Jesus Himself exercised his psychic abilities often during His three-year ministry here on earth. Besides being a mystic, he was truly a psychic and medium. He performed "miracles" that defied the laws of science and logic.

Are not these the gifts of the Holy Spirit that Christian believers speak about? Many Christian thinkers in fact speak of the seven gifts of the Holy Spirit (Crkivohlavek, Ken & Lorraine, 1976). Targ maintained that had it not been for the "positive results" of these miraculous works, "religion would have been abandoned centuries ago."

## Points of Convergence and Correlation

There are so many discoveries in the quantum world that unequivocally point to the existence of the psychic realm as another dimension. By delving deeper into the essence of reality, quantum physics has allowed us to better understand and appreciate the dynamics operating in the psychic realm. The correlation between the physical and psychical realm is becoming quite real and this is not only happening at the quantum level but also at the macro level.

In effect, quantum physicists seem to say to us that we human beings may also experience what is happening in the world of atoms, e.g., to tunnel through thick walls, jump from one place to another, be present in two or more places simultaneously and instantaneously or bi-locate. We can even travel forward or backward in time and back to the present, to levitate or float on water and air, to reincarnate or be born again.

Psychotherapist Lawrence Leshan argued that the importance of psychic occurrences is not so much that they are phenomena per se, but that they demonstrate the existence of that aspect of our being that must be lived in order for us to be truly human. He argued that if we continue to disregard these psychic events and corresponding psychic ability in us, then, we have not allowed a part of our identity to flourish and develop (see also Jennifer O'Neill, 2013).

The psychic realm, however, is not the be-all and end-all of our cosmic exploration. Mystics time and again have warned us that, while this stage of our journey may have to be passed and that we possess the innate capabilities to fully experience the psychic realm, clinging tenaciously to this realm could derail us away from the very purpose for which the Cosmos has set for us from the beginning (Leshan, 1974:47).

Indeed, many have been mesmerized and enthralled by their psychic powers to make it as their profession, not realizing that they have become trapped within this realm, forgetting that we are designed for something much higher in our cosmic journey. There is something more in us human beings than just be psychics or mediums, just as there is something in us that is not only physical and mental. We cannot live our life to the full by just being physically, mentally, and psychically developed.

# 8

## The Realm of Consciousness

*Only Consciousness exists—there is no existence apart from, or beyond, Consciousness. As Consciousness is All or One— It thus is not a higher Self, but the only Self.* - Peter Francis Cziuban

*We are not just in interaction with things, we are aware of them. It is through awareness that we are able to experience the world we interact with. We have access to being through awareness.* - David Kane

*Neuroscience….cannot explain where the aware self is located. On the other hand, if the self is distributed over a large area of the brain, what is it that binds this area together? What integrates all these different senses into one coherent and persisting sense of self?* - Deepak Chopra, et al.

*Of all the branches of science, neuroscience is the only one that has seriously challenged the dualistic view that the universe is divisible into matter and spirit. For at least a century, neuro-scientists have suspected that the machinery of the brain is somehow physically responsible for consciousness—for the soul itself.* - Michael S.A. Graciano

*Natural science does not simply describe and explain nature, it is part of the interplay between nature and ourselves. It describes nature as exposed to our method of question.* – Werner Heisenberg

*The discussion on Consciousness has always been a hot topic. The main concern revolves around the following questions: What is consciousness? Does it reside in our physical body, hence, a product of evolution? Or, is it something that came out of nowhere and merely embedded into our brain by some external forces? How do we know that we are conscious or not?*

*Over time, there emerged three views about Consciousness: the classical, quantum, and metaphysical. In the end, I highlight the new*

*development of a new consciousness paradigm that emerged from the synergy of these three. ###*

## The Classical Paradigm

Consciousness is not really one of the major concerns in classical science. During Newton's time, the term itself was almost non-existent. If it existed, then, this term was used to refer to those concepts that were more popular during those times, namely, mind, intelligence, alertness, memories, or intuition. In most cases, discussions on consciousness were avoided and its existence ignored. If forced to talk about it, the classical view would argue that consciousness simply emerged from the physical brain, reducible to the singularity that developed into more complex forms.

Following the Darwinian theory of evolution, molecular biologists today contend that after billions of years, the molecules gave birth to the primal cells that in turn metamorphosed into various body organs, including the brain out of which consciousness popped up. Biologist Edward O. Wilson (1998:221) remarked that "consciousness is…part of the system, intimately wired to all the neural and hormonal circuits regulating physiology." Neuroscientist Susan Greenfield (1998:213-214) likened it to "a light on a dimmer switch that grows as the brain does. The more complex the brain, the greater the consciousness."

Consciousness, then, is an emergent property, appearing as a result of what the brain does. Kak (2016) reechoed this materialist view when he said that man's awareness is a by-product of the complex neural events going on in the brain. In the classical thinking, consciousness losses its existence after the body dies (Richard M. Pico, 2002). As such, it is inconsequential, playing a minor role; it is not necessary for our existence and survival. Psychologist Julian Jaynes (1990:46-47) propounded this classical view saying that consciousness is not even needed in our thinking, in making judgments as well as in the functioning of our basic skills like speaking, writing, listening, or reading.

The classical view maintains that since it is emerging from the brain's various neural systems, "all we have to do is to find those parts of the brain that are responsible for consciousness, then trace out their anatomical evolution" (Julian Jaynes, 1990:16). And this is just what the mainstream sciences have been doing in the past few decades—tracing where in the brain does consciousness emerge and at what particular time in history did it appear. Rene Descartes argued that consciousness is situated in the pineal gland, a view that has been unchallenged for more than 200 years.

Today, some say it is in "that region referred to as the *upper brain-stem*, consisting largely of the thalamus and the midbrain," also known as the reticular formation since this is the most ancient part of the brain.

A more recent explanation argues that consciousness is closely associated with the "new brain" (the neocortex) since neuroscientists discover that it is in this region of the brain where intelligence, will, awareness, and other features that are believed to be attributable to consciousness, operates (Penrose, 1989).

With today's brain-imaging technologies, the problem of how consciousness arises can now be explained. According to Neuroscientist R. Joseph (2011), the visual input, for example, is transmitted from the eyes through the optic nerve to the midbrain and thalamus and is transferred to the primary visual receiving area in the neocortex of the occipital lobe. Once visual sensations reach the neocortex, the individual reaches that stage where he/she has consciousness of the outside world.

Below this area, we may not be aware at all of what various activities are going on within us (R. Joseph, 2011). Our pain sensation and response to it have to be brought upwards and transferred via the spinal cord to the neocortex, the central processing unit, in order to be aware of them. Voltaire Indigo (2015) explains this more picturesquely as follows:

*While reading this you probably forgot what the chair you're sitting in felt like and you probably are tuning out the sounds of the surrounding area and you are even blocking out the visual stimulus you receive from outside the pages of the book, you've forgotten what the room smells like and you aren't aware of the taste in your mouth.*

Some prominent evolutionists believe that the appearance of consciousness is sudden but were not so sure as to when it precisely appeared in our history. Lloyd Morgan (1923) simply stated that "Consciousness ... emerges as something genuinely new at a critical stage of evolutionary advance" and when it emerged it became the final threshold of the evolutionary process (Richard M. Pico, 2002). Later, when archeological findings began to be discovered, anthropologists argued that consciousness appeared during the time of the genus *Homo*, since this species already possessed the level of consciousness that we *Homo sapiens sapiens* have today.

But, on the whole, the classicalists are unable to fully explain the question of how physical entities like atoms and molecules work to produce the subjective state of consciousness. Thus, they would rather continue to disregard any discussion on consciousness because they feel they would be intruding into the world of subjective experiences, which is beyond the realm of physics. As Antti Revonsuo explained, hard-core sciences still "avoided consciousness or have been reluctant to put subjective experience into the focus of their research programmes" (2009).

During the time of Darwin, a minority view was already expressed. This view maintained that consciousness has its own existence apart from the physical body. Alfred Russell Wallace is quoted as saying that man's conscious faculties "could not possibly have been developed by means of the same laws which have determined the progressive development of the organic world in general, and also of man's physical organism" (Julian Jaynes, 1990:9). And as Julian Jaynes (ibid.) also defended: "There has to be more to human evolution than mere matter, chance, and survival. Something must be added from outside of this closed system to account for something so different as consciousness."

**The Quantum Paradigm**

The emergence of quantum physics in the 20th century offered a new explanation about the role of consciousness when it introduced into its classy mathematics the influence the conscious observer plays in creating and transforming reality. According to this view, it is the consciousness of the observer that creates reality and everything in creation is but a concrete expression of the wave nature of the original energy wave.

In quantum physics, consciousness has become primary and as such forms an essential part of reality. Without consciousness, the Cosmos and everything in it would not have appeared. In the "Interviews of Great Scientists" that appeared in *The Observer* on January 25, 1931, Max Planck (1858-1947) was quoted to express the same idea, as follows:

> *I regard consciousness as fundamental. I regard matter as derivative. We cannot get behind consciousness. Everything that we talk about, everything that we regard as existing, postulates consciousness.*

During the same interview, Erwin Schrödinger (1887-1961) was also quoted to have said:

> *Consciousness cannot be accounted for in physical terms. For consciousness is absolutely fundamental. It cannot be accounted for in terms of anything else.*

If this is so, then, we can also postulate that consciousness is as well very much active in the physical, mental, and psychic realms since these realms are simply its concrete manifestations in varying levels of frequencies. This correlates with Einstein's $E=mc2$. Energy is fundamental, not material in nature, responsible for the life and sustenance of everything that is in the Cosmos. Subhash Kak (2011) noted that consciousness has been there even before these cosmic realms came to exist:

> *Consciousness includes human mental process, but it is not just a human attribute. Existing outside space and time, it was 'there' 'before' those two words had any meaning.*

Consciousness becomes so fundamental that led Eugene Wigner to remark: "it was not possible to formulate the laws (of quantum theory) ... without reference to consciousness." Everything springs from Consciousness and, according to Kak (2016), to understand the Cosmos one needs to understand consciousness. Jean Houston (2016) even declared that: "Consciousness is the quantum field of the cosmos, the basic reality."

Since consciousness existed from the very beginning, it is but logical for quantum physicists to infer that consciousness has an existence of its own, not restricted by our dimensions of time, space, and matter. It is not a product of the mind, the mind being a derivative of the physical brain. According to Peter Francis Dziuban (2006:17), "The five forms of sensations, whether taken individually or as a group, never are *themselves* conscious … Sensations are merely unaware states of *reaction*," i.e., electro-chemical impulses of the various exchanges of atoms and molecules going on inside the neural systems of our brain. It cannot be attributed to as intelligence, thoughts, feelings, cognition, and other sensations.

Using purely rational and causality models, therefore, cannot explain what consciousness is all about. Hence, they are unable to reconcile the mental with the consciousness realm (Subhash Kak, 2016). There is no way how the subjective experience of awareness can be verified by experimentation using strictly the laws of physics and logic.

Consciousness belongs to a higher level of reality that envelops everything in the Cosmos, including us human beings. It is closely equivalent to what Leshan referred to as "superconscious" or "supraconscious" below which, according to him, lies the biological view of a conscious, subconscious, or unconscious mind.

In his words (1974:147): "The biological unconscious lies below the level of the conscious energies and the superconscious (genius, creative elan, the extra sensory, the divine inspiration, supraconscious intuition, etc.) lies above the level of any conscious, rational and logical thought or energy." In this view, Consciousness is different from the Freudian term "conscious mind," the latter being simply an attribute of the mind.

Our Consciousness is not a product of the complex processes going on in our brain. This is the reason why Consciousness is just beyond the mind to swallow hook-line-and-sinker for the mind is too limited to understand a much bigger entity (Larry Dossey, 2013). Because of this new perspective, quantum physics has opened the door to the possibility of expounding more about consciousness—its nature, origin, purpose— even if it clings to its materialist stance (Subhash Kak, 2016).

A series of controlled laboratory experiments confirmed this quantum view of consciousness. In the double-slit experiment, for instance, it was observed that it was because of the intrusion of the observer to investigate the behavior of electrons. In exploring reality, the observer carries with it his or her conscious attention and intention of what he or she wills to do. That subjective intention is not influenced by the operations and dynamics of atoms and molecules. On the contrary, it is this subjective interference that influences the latter (Henry Stapp).

Clinical studies in the medical sciences also confirmed that consciousness has its own existence independent of the body, even if the latter has been already declared "clinically dead." Testimonies of individual patients who have undergone near-death experiences (NDEs) assert that they were very much conscious of what was going on in the operating table. Erwin Laszlo (2016) "cites evidence from parapsychological research that there is communication 'beyond the reach of the eye and ear.'"

But like the classical thinking, the quantum view still maintains that consciousness is simply a product of what the brain does.

**The Metaphysical Paradigm**

The metaphysical view has been espoused by many world religions and mystical beliefs for thousands of years before modern science emerged.

Though their method and approach are different, their views resemble closely to that of quantum physics.

Michael Sharp (2006-2013) expounded a Kabbalistic cosmology where "consciousness is the root" of all things. Larry Dossey (2013) quoted a parallel view in the Hindu belief echoed by Ramakhrisna Rao that "Consciousness in the Indian tradition is more than an experience of awareness. It is a fundamental principle that underlies all knowing and being." Modern mystic Ahmed Hulusi (2011:20) correlated this fundamental principle or consciousness with the Pure Energy of modern science. In his words: "... the beginning of all life is pure energy; the point at which this energy transforms into matter is the atomic state, and matter in motion is basically the level of the body."

In the metaphysical view, Consciousness comes in several labels, e.g., ultimate reality, the ultimate source of everything, the giver and sustainer of life, the primal energy, or God. Sufi Ahmed Hulusi (2011) refers to it as the Absolute One, while Max Planck, in his famous lecture on "Religion and Science" in 1937, expressed that: "For believers, God is in the beginning, and for physicists He is at the end of all considerations… To the former He is the foundation, to the latter, the crown of the edifice of every generalized world view."

## The New Consciousness Paradigm

A new, all-encompassing scientific perspective of what Consciousness (with capital "C") is all about is now taking form. While the forerunners of mainstream science are still deeply rooted in classical physics, a few individuals have stood up, acknowledging that Consciousness could not just possibly be a product of evolution.

It has an existence of its own. It does not require the physical realm to exist and operate. On the contrary, it is the latter that depends on Consciousness for their life and existence. Even if the body and mind disappear, Consciousness remains. This in fact correlates with Einstein's $E=mc^2$. Energy is fundamental, non-material in nature, able to transform itself in varying levels of frequency. The diverse cosmic realms are simply frozen energy, manifested in various levels of vibration and frequency. If this is so, then, we can revise Einstein's formula to appear as $Consciousness=mc^2$.

In trying to describe what Consciousness is all about, it has become customary lately to distinguish between Consciousness as such and the "contents" or "stuff" of Consciousness (Shan Gao, 2014; Susan Greenfield, 1998:214). Consciousness per se is normally referred to as synonymous with awareness, wakefulness, or mindfulness (Roger Bartra, 2014; Stanislas Dehaene, 2014; Thich Nat Hanh, 1995; Eckhart Tolle, 2004).

The contents of Consciousness, on the other hand, refer to what we are "conscious of" since, as some authors would advance, whenever we are conscious, we are always conscious of something (Greenfield, 1998).

Laszlo (2016) calls this aspect of Consciousness "conscious property" to denote the same thing, "the ability of being conscious of something." Both the essence and the content of Consciousness, according to Shao, "form an external world independent of my mind."

I discovered three dominant contents of Consciousness, namely, Self-Consciousness, Cosmic Consciousness, and Divine Consciousness. Each of these categories is neither separate nor exclusive; they can manifest themselves simultaneously at varying degrees.

The Self is the first direct experience of the individual. It can be expressed in the form of awareness to our: bodily sensations (e.g., pains, numbness, breathing, pulse, heart beating, temperature, etc.); mental states (thoughts, emotions, desires, etc.); natural surroundings (e.g., insects, plants, animals, caves, lakes, etc.); psi abilities (e.g., telepathy, aura sensitivity, remote viewing, etc.); our own being, usually referred to as the Self, I Am, or Soul; and awareness of our oneness and communion with the Divine, the Matrix of Everything.

According to Adams Zeman (2003), *Self-Consciousness* can be both sensory and perceptual and can include: "the family of bodily sensations, tingles, itches, aches and pains and the deliverances of the five traditional senses: all that we see, hear, taste, smell, touch." But it is not only our bodily sensations.

Richard Pico (2001:309) maintains that Self-Consciousness is likewise awareness of our perceptions, thoughts, and emotions. It is awareness of what is going on in our mind like those subjective bodily sensations of joy, happiness, anguish, suffering, worry, depression, and other mental states.

Mystics and yogis refer to the Self as the one that watches and observes the bodily sensations, streams of thoughts and emotions, as well as incidences of psi abilities that appear and disappear every now and then.

Beyond the realm of *Self-Consciousness*, there lies an even bigger experience of awareness that embraces the Cosmos as a whole and our intimate relationship with everything in it. I refer to this level of awareness as *Cosmic Consciousness*.

This level of Consciousness includes our awareness of the five elements of nature, namely, air, fire, earth, water, space and the role the play in our life. It includes awareness of our close connection with subhuman species and the role they play in the entire cosmic scheme (Richard Maurice Bucke, M.D., 2011).

David Kane (2013) expressed that: "We are not just in interaction with things, we are aware of them. It is through awareness that we are able to experience the world we interact with. We have access to being through awareness."

In the real world, modern physics shows that our Consciousness connects us quite intimately with every living and non-living things. Nobel Physicist Erwin Schrödinger described our profound interconnectedness this way: "Consciousness is a singular of which the plural is union. There is only one thing, and that which seems to be a plurality is merely a series of different aspects of this one thing, produced by a deception, the Indian maya, as in a gallery of mirrors."

*Cosmic Consciousness* is very much parallel to David Bohm's concept of unbroken wholeness, in which every part of the Cosmos is intimately connected with each other, not as isolated parts and pieces, but as one whole system, each of which is necessary for the functioning of the whole. Consciousness is in everything, it complements with everything in the Cosmos in varying degrees and attributes. As Ali Nomad (2014:24) explained: "It harmonizes with and blends into all the various degrees and qualities of consciousness in the cosmos, and becomes "at-one" with the universal heart-throb."

But beyond *Self* and *Cosmic Consciousness* is the third and highest level, namely, the *Divine Consciousness*. This level of Consciousness refers to our awareness of the realm of the spirits, the discarnate entities, the Unmanifested, uncreated, and unborn. Goswami et al. (1993) described this as the awareness of a Being that is ascribed to as the "original, self-contained, and constitutive of all things."

Quantum physicists do not dispute or disprove the existence of the Divine or God. Werner Heisenberg, for example, exclaimed that: "The first gulp from the glass of natural sciences will turn you into an atheist, but at the bottom of the glass God is waiting for you."

They in fact personally ventured into studying the world's greatest religions. Its implicit reference to a creator is now mounting, referring to it as the "Cosmic Mind and Intelligence" who is also the" Matrix of All Things."

Humanity's evolutionary process is an ascent from the physical to the metaphysical. In particular, Science has pushed the cosmic evolutionary process as a progressive ascent towards something higher than the physical, mental, psychic, and Consciousness realms. As Max Planck advanced:

> *Science...means unresting endeavor and continually progressing development toward an aim which the poetic intuition may apprehend, but the intellect can never fully grasp.*

Ali Nomad (2014:8) noted the beliefs of Religion and Mysticism that Self and Cosmic Consciousness ultimately lead us to the awareness and realization of our Higher Self, which culminates only when we are finally "at-one-ment with the Om."

Nevertheless, from the perspective of quantum physics, the issue of God-consciousness remains unsettling. While Religion has always taught us to believe in God who created everything, quantum physics has avoided any discussion about God since it is a subject that goes beyond the physical. It believes that the Cosmos does not need any creator at all to realize its existence. If I may paraphrase Carl Sagan, in his video presentation "The Cosmos," he raised a rhetorical question: "If God always existed, as Religion and Mysticism avow, then, what hinders us scientists from also concluding that the Cosmos always existed and that there's no need of a creation."

If this is so, then, it is in the Consciousness realm that we humans differ with the chimpanzees. It is in this realm that we cross the threshold of the physical and enter the world of the metaphysical. We have one of our feet set on the physical and the other on the metaphysical. We stand on the realm where the body and the spirit meet, where reason and intuition interface.

# In the Spring of Life: The Dawning of an Awakened Humanity

*When I consider thy heaven, the work of thy fingers, the moon and stars, which thou hast ordained; what is mankind that you are mindful of them, human beings that you care for them?* - Psalm 8:3-4

*I consider science an integrating part of our endeavor to answer the one great philosophical question which embraces all others – who are we? And more than that: I consider this not only one of the tasks, but the task, of science, the only one that really counts.* - Erwin Schrödinger

*We are the cosmos made conscious and life is the means by which the universe understands itself.* - Brian Cox

*I am a piece of existence, but I am also whole. But I am also so constructed that I cannot be surveyed at single glance.* - Erwin Schrödinger

*I have inside me the winds, the deserts, the oceans, the stars, and everything created in the universe.* - Paulo Coelho

*As Julius Caesar crossed the Rubicon, a new, enlightened humanity emerged.*

*As the story goes, Caesar suddenly halted his troops on the banks of the river Rubicon. He was in a deep dilemma whether to pursue his enemies or not. He turned to his troops and said,*

*"We may still retreat; but if we pass this little bridge, nothing is left for us but to fight it out in arms."*

*Unable to decide immediately, the cosmic forces intervened. This happened when a man from out of nowhere with trumpet on hand, blew it off with an electrifying blast, sounding the advance. Then, he crossed to the other side, instantly making up Caesar's mind, jolting him to exclaim:*

*"Let us go whither the omens of the gods and the inequity of our enemies call us. THE DIE IS CAST."*

*As Caesar crossed the Rubicon, he changed the life and elevated the status of humanity to a higher dimension. Everybody sprung to life, and along with it the emergence of a more animated and enlightened humanity. Soldiers became ecstatic, infused with more courage and vigor, a strong bond of unity and connectivity as well.*

*The mysterious man was never seen again. Nobody knew his name; he was very casual when he appeared, acting like an accident but he was no accident.*

*We are reminded of Arjuna who refused to fight against his brothers on the other side of the battlefront. But again the Cosmos intervened. A terrifying force surged from behind, encouraging him to advance and fight in a bloody carnage. Like Julius Caesar, Arjuna immediately raised humanity to a higher level of existence.*

*We are living in a conscious universe that every now and then sends us somebody that jerks us to action. Then, as suddenly as this "mysterious somebody" appears before us, everything in us springs to life again, making us more fully conscious and invigorated to confront the daily battle we face in life. Then, this "somebody" abruptly leaves us just as mysteriously as he appeared.*

*The Cosmos we are living in is alive and conscious. It is constantly communicating with us, churning out messages that still need to be decoded. Yet, in the end, it's always reminding us that we are on an ascent to a higher realm of existence.* ###

Spring has begun and I am beginning to smell the fresh scents of a new life. Flowers are starting to bloom, insects beginning to crawl the earth, and the birds gleefully making their presence in the skies, giving a kaleidoscope of colors and patterns against a celestial backdrop that is slowly yielding to unclouded and sunny days ahead. The hills and rivers are vibrant, teeming with life. Everyone is in a celebratory mode enjoying Nature's abundance. Gone are the fears and worries experienced in the past.

Soon summer will come, a time to reap and festively enjoy the fruits of our hard labor. Paradoxically, not all of us will be able to savor and relish its abundance. Many will continue to wallow in need and poverty, unable to access and partake of Nature's abundance. Some will recover from their health issues but others will continue to suffer in pain. In spite of all this, we will face summer undaunted. Life must go on.

But autumn will soon take over, a time when Nature's abundance will begin to dwindle and become scarce. We will again experience what it means to be in need, what it means to be emotionally and physically down. Eventually, we will be back to the winter season when abundance will disappear and become dormant and dead.

Such is the cyclical pattern of time and life the Cosmos has fashioned for us. Spring signals the appearance of a new life and with it the dawning of a vibrant and an awakened humanity. Going from one season to another, we will remain fearless. As in the past, we will emerge unscathed as new and spirited individuals. The past has left us a legacy, a wealth of experience and precious lessons that guide us today every moment of our life. We will continue to move on.

The lessons we learned encouraged us to expand our knowledge, sharpen our skills, invent sophisticated technologies, and organize ourselves into groups to better cope with any eventuality. We established norms of conduct and laws for our respective units to function harmoniously as a people. Over the millennia, our accumulated beliefs strengthened us that we later institutionalized these formally as religions to accommodate our faith in the supernatural.

We learned our lessons and we prepared ourselves to brace for whatever possibilities the future brought us. We were awed of the future's uncertainty but at the same time we discovered our dormant capabilities we once thought we don't have and learned to unleash these talents to the fullest. We are not automatons, after all.

Spring marks the birthing of a new life and along with it the dawning of a conscious and vibrant humanity that infuses in us a renewed vigor and strength to face the challenges expected of us humans, hailed as the zenith of creation. What the future may still bring and how we will respond to it will depend on what we intend and will today.

New events still await us. But along with it is the birthing of a new, expanding view of ourselves. Who we might be and what our role will be in the entire cosmic arrangement and progression remains to be structured by us. We know that our preparation is not an assurance of the kind of future we will envision. But we will "keep on keeping on," as our class slogan when I was in Grade VI proclaimed.

Life is indeed an unending process of education and learning our lessons. Life ceases when we stop educating ourselves. Physicist Isaac Asimov could be right: "Education is not something you can finish." Exploration into the unknown continues. What may become of us still remains to be seen. Each moment the clock ticks, the Cosmos always discloses something new about itself and our place in it. It makes itself intelligible to us but, in mysterious ways, churns out messages that still need to be deciphered and decoded.

As I end this book this springtime, I spring to life again, wanting to start another project. I am inviting the readers to join me in this spring of life again. I don't promise anything except the prospect of discovering new in the world of uncertainty and the unknown.

Let me end here with a quote from Carl Sagan: "Somewhere, something incredible is waiting to be known."

# References

Adams, Fred. 2002. *Origins of Existence: How Life Emerged in the Universe*. New York: The Free Press.

A. Nicholas Frank. 1996. *SocioEconomic Synergism: Synergistic Individual, Organizational and Community Renewal*. L.A., California: BioFitness Center.

Aïvanhov, Omraam Mikhaël. 2011. *Cosmic Balance: The Secret of Polarity*. Izvor Collection – No. 237. Prosveta. Translated from French. Amazon Digital Services, Inc.

Al-Khalili. Jim. 2012. *Quantum: A Guide for the Perplexed*. Weidenfeld & Nicolson. Amazon Digital Services, Inc. 288 pages.

Amthor, Frank. 2011. *Neuroscience for Dummies*. Amazon Digital Services, Inc. 384 pages.

Anonymous. 2009. *The Cloud of the Unknowing*. London. Amazon Digital Services, Inc. with an Introduction by Evelyn Underhill. A new translation by Carmen Acevedo Butcher appears in 2011. Boston & London: Shambala. Amazon Digital Services, Inc. 307 pages.

Asimov, Isaac. 1992. *Atom: Journey Across the Subatomic Cosmos*. New York: Penguin Books USA Inc.

Atkinson, William Walker & Lateef Terrell Warnick. 2012. *Yeshua: Mystic Christianity & The Path to Christ Consciousness*. USA

Aurobindo, Sri. 1957. *The Synthesis of Yoga.* Pondicherry, India: Aurobindo Ashram Press.

Bailey, Alice. A. 1993. *The Rays of Initiations*. Lucis Publishing Company. 820 pages.

Barnett, H.G. 1972. "Anthropology as an Applied Science." Jesse D. Jennings and E. Adamson Hoebel. *Readings in Anthropology*. New York: McGraw-Hill Book Company. Pp. 482-486.

Bartra, Roger, Roger. 2014. *Anthropology of the Brain: Consciousness, Culture, and Free Will.* Cambridge: University Printing

House. Amazon Digital Services, Inc. 208 pages. Translated by Gusti Gould.

Barrow, John D. 1994. *The Origin of the Universe*. New York: Basic Books.

Barton, George A. 1937 *Archaeology and The Bible,* 7th Edition revised, (Philadelphia: American Sunday School, 1937), pg. 296. This can now accessed for the public at http://www.piney.com/BabEnumFragAnim.html

Behe, Michael J. 2007 *The Edge of Evolution: The Search for the Limits of Darwinism*. New York: Free Press.

_____. 1998. "Intelligent Design Theory as a Tool for Analyzing Biochemical Biochemical Systems." In Dembski, William A. (ed.). *Mere Creation: Science, Faith & Intelligent Design*. Downers Grove, Illinois: InterVarsity Press. Pp. 177-194.

Bennett, C. H., G. Brassard, C. Crépeau, R. Jozsa, A. Peres, W. K. Wootters. 1993. "Teleporting an Unknown Quantum State via Dual Classical and Einstein-Podolsky-Rosen Channels." *Phys. Rev. Lett.* **70**, 1895-1899.

Benyus, Janine M. 2009. *Biomimicry*. HarperCollins Publishers. Amazon Digital Services, Inc., 324 pages.

Besant, Annie. 1994. *Theosophical Society of the Philippines Digest*, 1994, 4th qrtr., January 7, 2012, pp. 82-87.

Black, J.A., G. Cunnigham, E. Robson, and G. Zlyyomi. 1998. *Cattle and the Grain*. Oxford University Press. Their electronic Homepage can be accessed at http://etcsl.orinst.ox.ac.uk/#. See also "How Grain Came to Sumer" at http:// www.piney.com/BabGrainSum.html.

Bohm, David. 2012. *Quantum Theory (Dover Books on Physics)*. Dover Publications. Amazon Digital Services, Inc. 655 pages.

_____. 2012(b). *On Creativity*. New York: Routledge. Amazon Digital Services, Inc. 188 pages. First published in 1966 by Routledge. With a new preface by Leroy Little Bear. Edited by Lee Nichol.

_____. 2005. *Wholeness and the Implicate Order*. Routledge.

Amazon Digital Services, Inc. 305 pages. First published in 1980 by Routledge and Kegan Paul, New York.

_____. 1980. *Wholeness and the Implicate Order.* Routledge & Kegan Paul.

Bohm, D. & B. Hiley. 1975. "On the Intuitive Understanding of Nonlocality as Implied by Quantum Theory," *Foundations of Physics,* vol. 5 (1975).

Bohr, Neils. 1958. *Atomic Physics.* New York and London: John Wiley and Sons.

Brenan, Barbara Ann. 2011. *Light Emerging: The Journey of Personal Healing.* Amazon Digital Services, Inc. 352 pages.

Brooks, Michael (ed.). 2016. "The Weirdest of the Weird." *New Scientist: The Collection Book 3. The Quantum World: Your Ultimate Guide to Reality's True Strangeness.* Amazon Digital Services LLC. 238 pages.

Brooks, Rodney A. 2010. *Fields of Color: The Theory That Escaped Einstein.* Epic Publications. Amazon Digital Services, Inc. 175 pages.

Brown, Jonathon. 1998. *The Self.* New York: Routledge. Amazon Digital Services, Inc.

Bucke, Richard Maurice, M.D. 2011. *Cosmic Consciousness: A Study in the Evolution of the Human Mind.* Published by MCMI. Amazon Digital Services, Inc. 517 pages.

Campbell, Joseph. 1972. *The Hero With A Thousand Faces.* Princeton University Press. 416 pages.

Capra, Fritjof. 2000. *The Tao of Physics: An Exploration of the Parallels Between Modern Physics and Eastern Mysticism.* Boston: Shambala.

_____. 1975. *The Tao of Physics: An Exploration of the Parallels Between Modern Physics and Eastern Mysticism.* Boston: Bantam Books, Shambala Publications, Inc.

Carter, Chris. 2010. *Science and the Near-Death Experience: How Consciousness Survives Death.* Inner Traditions. Amazon Digital Services, Inc. 320 pages.

Chevalier, William J. 2011. *A Short History of Time, Space and the Quantum World: A Non-Scientist's Guide to Uncertainty, Teleportation, Digitization, Consciousness and Time Travel*. Australia. Amazon Kindle edition.

Cori, Patricia. 2010. *The Cosmos of the Soul: A Wake-Up Call for Humanity*. Berkeley, California: North Atlantic Books. Amazon Digital Services, Inc., 274 pages.

Charles, R.H. (translator). 1906. *The Book of Enoch*. Translated from the Ethiopian. English E-text edition scanned by Joshua Williams, Northwest Nazarene College, 1995. Edited by Wolf Carnahan, 1997.

Chardin, Teilhard de. 1961. *The Phenomenon of Man*. Harper Torchbooks, The Cloister Library, Harper & Row, Publishers

Copleston, Frederick. 1977. *A History of Philosophy: Maine de Biran to Sartre*. Garden City, New York: Image Books. Volume 9, Part II: Bergson to Sartre.

Cornell, James (ed.). 1990. Quoting the *Principia*, General Scholium, translated in the *Principia*, Berkeley: University of California Press, 1962, vol. 2.

Cox, Brian & Jeff Forshaw. 2012. *The Quantum Universe (and Why Anything That Can Happen Does Happen)*. Da Capo Press. Amazon Digital Services, Inc. 272 pages.

Cremo, Michael A. 2011. *Human Devolution: A Vedic Alternative to Darwin's Theory.*Amazon Digital Services, Inc. Originally published in 2003 by Torchlight Publishing. 554 pages.

Crick, Francis. 1982. *Life Itself: Its Origin and Nature*. Macdonald. Simon & Schuster.

Crumley, Carole L. 2001. *New Directions in Anthropology & Environment: Intersections*. Lanham, Maryland: Rowman & Littlefield Publishers, Inc.

Damasio, Antonio. 2010. *Self Comes to Mind: Constructing the Conscious Brain*. Vintage. Random House LLC. Amazon Digital Services, Inc. 384 pages.

Danielson, Dennis Richard (ed.). 2000. *The Book of the Cosmos:*

*Imagining the Universe from Heraclitus to Hawking.* Cambridge, Massachusetts: Perseus Publishing.

Davies, Paul. 1988. *The Cosmic Blueprint: New Discoveries in Nature's Creative Ability to Order the Universe.* New York: Simon and Schuster.

_____. 1983. *God and the New Physics.* New York: Simon & Schuster, Inc.

Davies, Paul and John Gribbin. 1992. *The Matter Myth: Dramatic Discoveries That Challenge Our Understanding of Physical Reality.* New York: Simon & Schuster.

Dawkins, Richard. 2006. *The Selfish Gene: 30th Anniversary Edition.* Oxford University Press. 384 pages.

De Broglie, Prince Louis. 1970. *The Revolution in Physics: A Non-mathematical Survey of Quanta.* Greenwood Press. 310 pages.

Dehaene, Stanislas. 2014. *Consciousness and the Brain: Deciphering How the Brain Codes Our Thoughts.* Penguin Books. Penguin Group (USA) LLC. Amazon Digital Services, Inc. 352 pages.

Déli, Eva. 2015. *The Science of Consciousness: How a New Understanding of Space and Time Infers the Evolution of the Mind.* Amazon Digital Services, Inc. 222 pages.

Deutsch, David. 1998. *The Fabric of Reality: The Science of Parallel Universes—and Its Implications.* USA: Penguin Books. Amazon Digital Services, Inc. 404 pages.

Denbigh, K.G. 1975. *An Inventive Universe.* London: Hutchinson & Co.

Di Chardin, Teilhard. 1976. *The Phenomenon of Man.* Harper Perennial.

Di Corpo, Ulisse and Antonellas Vannini. 2014. *The Balancing Role of Entropy/Syntropy in Living and Self-Organizing Systems: Quantum Paradigm.* Amazon Digital Services, Inc. 248 pages.

Dispenza, Joe. 2007. *Evolve Your Brain: The Science of Changing Your Mind.* FL: Health Communications, Inc. Amazon Digital Services, Inc. 508 pages.

Dishit, Sudhakar S. 1989. *I am All: Cosmic Vision of Man*. Published by Chetana. 172 pages

Dixon, Macneile A. 2013. *The Human Situation*. Reitell Press. Amazon Digital Services, Inc. 448 pages.

Dossey, Larry. 2013. *One Mind: How Our Individual Mind is Part of a Greater Consciousness and Why It Matters*. New York: Hay House, Inc.

Dunbar, Roman. 2014. *Human Evolution: A Pelican Introduction*. Penguin Books. Amazon Digital Services, Inc. 388 pages.

Duncan, Anthony. 2012. *The Conceptual Framework of Quantum Field Theory*. Oxford: Oxford University Press. Amazon Digital Services, Inc. 768 pages.

Durant, Will. 2011. *Our Oriental Heritage: The Story of Civilization*. Simon & Schuster: Simon and Schuster Digital Sales Inc. Volume I. 1,049 pages.

Dyson, Freeman. 1979. *Disturbing the Universe*. New York: Harper & Row, Publishers.

Dziuban, Peter Francis. 2006. *CONSCIOUSNESS is ALL: Now Life is Completely New*. Nevada City, CA: Blue Dolphin Publishing, Inc. Amazon Digital Services, Inc. 31 pages.

Eccles, Sir John. 2013. *Mind and Brain: The Many-Faceted Problems*. Paragon House. Amazon Digital Services, Inc. 300 pages.

Eliade, Mircea. 1977. *From Primitives to Zen: A Thematic Sourcebook of the History of Religions*. San Francisco: Harper & Row, Publishers.

_____. 1954. *The Myth of the Eternal Return or, Cosmos and History*. Princeton, N.J.: Princeton University Press. Translated from the French by Willard R. Trask.

Erway, Daniel (aka Nirmala). 2011. *That is That: Essays About True Nature*. Endless Satsang Foundation. Amazon Digital Services, Inc. 172 pages.

Falk, Dan. 2016. "New Support for Alternative Quantum View." *Quanta Magazine*. May 16, 2016. Can be accessed at https://www.

quantamagazine.org/20160517-pilot-wave-theory-gains-experimental-support/.

Finlay, Graeme. 2013. *Human Evolution: Genes, Genealogies and Phylogenies*. New York: Cambridge University Press. Amazon Digital Services, Inc. 368 pages.

Flammarion, Camille. 1907. *Mysterious Psychic Forces: An Account of the Author's Investigations in Psychical Research, Together with those of other European Savants*. Boston: Small, Maynard and Company and Cambridge: The University Press.

Fleming, Ray. 2012. *The Zero-Point Universe*. Amazon Digital Services, Inc.307 pages.

Follett, Scott. 2015. *Quantum Beginning*. Amazon Digital Services, Inc. 110 pages.

Forbes. Nancy and Basil Mahon. 2014. *Faraday, Maxwell, and the Electromagnetic Field: How Two Men Revolutionized Physics*. Prometheus Books. 320 pages. Amazon Digital Services, Inc.

Freese, Kathrine. 2014. *The Cosmic Cocktail: Three Parts Dark Matter: Science Essentials*. Princeton University Press. Amazon Digital Services, Inc. 263 pages.

Gamow, George. 1972. *One Two Three...Infinity*. New York: The Viking Press. Inc. 8th printing.

Gao, Shan. 2014. *Dark Energy: From Einstein's Biggest Blunder to the Holographic Universe*. Kindle Direct Publishing. 33 pages.

Gazzaniga, Michael. 2013. *The Biology of the Mind*. Amazon Digital Services, Inc.

_____. 2011. *Who's in Charge?: Free Will and the Science of the Brain*. HarperCollins Phublishers. Amazon Digital Services, Inc. 275 pages.

_____. 1998. *The Mind's Past*. Berkely and Los Angeles, California: California University Press.

Gazzaniga, Michael S., Richard B. Ivry., George R. Mangun. 2013. *Cognitive Neuroscience: The Biology of the Mind*. Amazon Digital Services, Inc., 752 pages.

Ghare, Madhavi. 2007. *Paranormal Activity*. Also in http:// www. buzzle.com/articles/ paranormal-activity.html.

Gleiser, Marcelo. 1997. *The Dancing Universe: From Creation Myths to the Big Bang*. New York: A Dutton Book.

Goldman, I. 2000. *Micro to Macro: Policies and Institutions for Empowering the Rural Poor*. South Africa: Kenya.

Govinda, Lama Anagarika. 1974. *Foundations of Tibetan Mysticism*. New York: Samuel Weiser.

Greene, Brian. 2011. *The Hidden Reality: Parallel Universes and the Deep Laws of the Cosmos.* Vintage. Amazon Digital Services, Inc. 464 pages.

_____. 2007. *The Fabric of the Cosmos: Space, Time, and the Texture of Reality.* Vintage. Random House LLC. Amazon Kindle Direct Publishing. 592 pages.

_____. 1999. *The Elegant Universe: Superstrings, Hidden Dimensions, and the Quest for the Ultimate Theory.* N.Y.: W.W. Norton & Company.

Greenfield, Susan. 1998. "How Might the Brain Generate Consciousness?" In Rose, Steven (ed.). *From Brains to Consciousness?: Essays on the New Sciences of the Mind*. Princeton, New Jersey: Princeton University Press. Pp. 210-227.

Greenstein, George. 1988. *The Symbiotic Universe: Life and Mind in the Cosmos*. New York: William Morrow.

Guterl, Fred. 2013. "Introduction: The Sprawling Story of Human Evolution." In *Scientific American Editors. Becoming Humans: Our Past, Present and Future*. Amazon Digital Services Inc. 189 pages.

Guth, Alan H. 1997. *The Inflationary Universe: The Quest for a New Theory of Cosmic Origins*. New York: Addison-Wesley Publishing Company, Inc.

Harari, Yuval Noah. 2015. *Sapiens: A Brief History of Humankind.* Amazon Digital Services, Inc. 465 pages.

Hardesty, Donald L. and Don D. Fowler. 2001. "Archaeology and Environmental Changes." In Crumley, Carole L. 2001. *New Directions in Anthropology & Environment: Intersections.* Lanham, Maryland: Rowman & Littlefield Publishers, Inc. Pp. 72-89.

Harris, Edward G. 2014. *A Pedestrian Approach to Quantum Field Theory.* Dover Publications. Amazon Digital Services, Inc. 192 pages.

Harth Erich. 1995. *The Creative Loop: How the Brain Makes a Mind.* New York: Addison-Wesley Publishing Company. First paperback printing.

Hatfield, Brian. 2014. *Quantum Field Theory of Point Particles and Strings (Frontiers in Physics).* Westview Press. Amazon Digital Services, Inc. 752 pages.

Haviland, William A. 2000. *Anthropology.* Orlando, FL: Harcourt College Publisher.

Heidel, Alexander. 1963. *The Gilgamesh Epic and Old Testament Parallels.* University of Chicago Press. 269 pages. Second edition. The first edition appeared in 1951.

Heisenberg, Werner. 1990. *Across the Frontiers.* Woodbridge. CT: Ox Bow Press.

_____. 1958. *Physics and Philosophy.* New York: Harper Torchbooks.

Hiley, Basil and F. David Peat. 2012. *Quantum Implications: Essays in Honour of David Bohm.* Routledge. Amazon Digital Services, Inc. 462 pages.

Holland, Gail Bernice. 1998. *A Call for Connection: Solutions for Creating a Whole New Culture.* Canada: Publishers Group West.

Holmes, Lowell D. 1971. *Anthropology: An Introduction.* New York: The Ronal Press Company.

Holzner, Steven. 2013. *Quantum Physics For Dummies, Revised Edition.* New Jersey: John Wiley & Sons, Inc. Sold by Amazon Digital Services, Inc. 336 pages.

Hood, Bruce. 2012. *The Self Illusion: How the Social Brain Creates Identity.* Oxford University Press, USA. Amazon Digital Services, Inc. 365 pages.

Hua-Ching Ni (translator). 2003. *The Complete Works of Lao Tzu: Ching, Tao The & Hua Hu Ching.* Santa Monica: Seven Star Communications. 12th printing.

Hua-Ching Ni. 2003. The description of the *yin* and *yang* is lifted from Marcelo Gleiser (1997:18).

Hulusi, Ahmed. 2011. *Universal Mysteries.* Istanbul. Amazon Digital Services, Inc.

Itzykson and Jean-Bernard Zuber. 1980. *Quantum Field Theory.* New York: Dover Publications, Inc. Also released in 2012 by Amazon Digital Services, Inc., 752 pages.

Jacklewski, Tom. 2011. *Zero Point Energy Field Effects: The Physics of Free Energy.* Amazon Digital Services, Inc., 111 pages.

Jacobsen, Thorkild. 1987. *The Harps that Once . . . Sumerian Poetry in Translation.* New Haven and London: Yale University Press.

James, William. 2012. *Varieties of Religious Experience, a Study in Human Nature.* Amazon Digital Services, Inc. 234 pages.

Jones, Andrew Zimmerman and Daniel Robbins. 2009. *String Theory for Dummies.* For Dummies. Amazon Digital Services, Inc., 384

Joseph, Rhawn. 2001. *Astrobiology, the Origin of Life and the Death of Darwinism.* San Jose, California: University Press California.

Kafatos, Menas and Robert Nadeau. 1990. *The Conscious Universe: Parts and Wholes in Physical Reality.* Springer. Amazon Digital Services, Inc. 183 pages.

Kak, Subhash. Ph.D. (editor of Volume 3). 2011. "Quantum Physics & Neuroscience of Mind." *Journal of Cosmology*).

Kaku, Michio. 2014. *The Future of the Mind: The Scientific Quest to Understand, Enhance, and Empower the Mind.* Doubleday. Random House LLC. Amazon Digital Services, Inc. 400 pages.

_____. 1998. *Introduction to Superstrings and M-Theory.* Springer. Amazon Digital Services, Inc. 609 pages.

_____. 1997. *Visions: How Science Will Revolutionize the*

*21st Century*. New York: Anchor Books. Amazon Digital Services, Inc., 418 pages.

Kane, David. 2013. *Awareness and Being*. Amazon Digital Services, Inc. 66 pages.

Khroutski, Konstantin S. 2001. "Bringing Forward the Philosophy of Universal Science: A Cosmist Concept." Can be accessed at http://nb.vse.cz/kfil/elogos/science/ khrout1-03.pdf.

King, Leonard W. 1902. *The Seven Tablets of Creation*. Published a fuller text of the *Epic of Creation or Enuma elish*. The complete text can also be accessed at http:// www.sacred-texts.com/ane/stc/index. htm.

Kramer, Samuel Noah. 1981. *History Begins at Sumer*. Philadelphia: University of Pennsylvania Press. The original edition was published in 1959 (Garden City, N.Y.: Doubleday/ Anchor.

_____. 1961. *Sumerian Mythology*, Harper & Brothers, New York, 1961. See also the revised edition by Philadelphia: U of Pennsylvania P, 1972.

_____. 1961(a). *Sumerian Mythology: A Study of Spiritual and Literary Achievement in the Third Millennium B.C.*, Harper & Brothers, New York, 1961. See also the revised edition by Philadelphia: U of Pennsylvania P. 1972.

Krebs, Robert E. 2003. *The Basics of Earth Science*. Westport, Connecticut: Greenwood Press.

Krivohlavek, Ken & Lorraine. 2004. *Desire Spiritual Gifts*. Philippines: ICI Ministries. 184 pages.

Kumar, Manjit. 2008. *Quantum: Einstein, Bohr and the Great Debate About the Nature of Reality*. Icon Books Ltd. Amazon Digital Services, Inc. 480 pages.

Küng, Hans. 1988. *Theology for the Third Millennium*. New York: Doubleday. Translated by Peter Heinegg

Kurtz, Paul (ed.). 2003. *Science and Religion: Are They Compatible?* New York: Prometheus Books.

Larson, Cynthia Sue. 2015. "The Art of Quantum Jumping: How to

Shift Your Reality in Big, Positive Ways." Accessed in 2015 at http://www.consciouslifestylemag.com/quantum-jumping-shift-reality/. This article is excerpted from her other article titled "Quantum Jumps: An Extraordinary Science of Happiness and Prosperity." http://www.realityshifters.com/pages/realityshifts.html.

_____. 2013. *Quantum Jumps*. Amazon Digital Services, Inc. 284 pages.

_____. 2012. *Reality Shifts: When Consciousness Changes the Physical World*. USA: Amazon Kindle Direct Publishing. 310 pages.

Laszlo, Ervin and Anthony Peake. 2014. *The Immortal Mind: Science and the Continuity of Consciousness beyond the Brain*. Inner Traditions. Amazon Digital Services, Inc. 176 pages.

Le Bihan, Denis. 2014. *Looking Inside the Brain: The Power of Neuroimaging*. Princeton University Press. Amazon Digital Services, Inc. 168 pages.

Lederman, Leon. 2012. *The God Particle: If the Universe is the Answer, What Is the Question*. Mariner Books. Houghton Mifflin Harcourt. Amazon Digital Services, Inc. 448 pages.

Le Grice, Keiron. 2011. *The Archetypal Cosmos*. Floris Books. Amazon Digital Services, Inc. 328 pages.

Leibniz Gottfried. 1670. *Philosophical Writings, Philosophical Writings*. 7 vols., 1663–90.

LeMind, Anna. 2016. "Parallel Worlds Exist And Interact With Each Other." Posted by sunshyne49 on September 14, 2016.

Leshan, Lawrence. 1974. *How to Meditate: A Guide to Self-Discovery*. New York: Bantam Books. 161 pages. See Chapter 7 on "Mysticism, Meditation and the Paranormal," pp. 47-51.

Levy, David H. 2003. *Cosmology 101*. New York: Simon & Schuster.

Lincoln, Dan. 2005. *Understanding the Universe: From Quarks to the Cosmos*. New Jersey: World Scientific.

Lorentz, Henrik Antoon. 2009. *The Einstein Theory of Relativity:*

*An Explanation and Appreciation.* Originally published 1920.

Lovegren, Stefan. 2004. "Teleportation Takes Quantum Leap." (http://news.nationalgeographic.com/news/2004/08/0818_040818_teleportation_2.html).

Lowis, Steven. H. 2011. *The Meaning of Life: Understanding Purpose and the Nature of Reality.* Amazon Digital Services, Inc. 201 pages.

Luckey, John. 2015. *In Search of a Quantum God.* Amazon Digital Services. 382 pages.

Ma, X. S.; Herbst, T.; Scheidl, T.; Wang, D.; Kropatschek, S.; Naylor, W.; Wittmann, B.; Mech, A.; et al. (2012). "Quantum teleportation over 143 kilometres using active feed-forward". *Nature.* 489 (7415): 269–273.

Mandi, Franz and Graham Shaw. 2013. *Quantum Field Theory.* Wiley. Amazon Digital Services, Inc. 492 pages.

Maudlin, Tim. 2011. *Quantum Non-Locality and Relativity: Metaphysical Intimations of Modern Physics (Aristotelian Society Monographs.* Wiley-Blackwell. Amazon Digital Services, Inc., 312 pages.

McCabe, Joseph. 2014. *The Evolution of Civilization.* London: Watts & Co. Amazon Digital Services, Inc. 150 pages.

McConnell, Craig Sean. 2002. "Twentieth-Century Cosmologies." In Ferngren, Gary B. (ed.). 2002. *Science and Religion: A Historical Introduction.* Baltimore and London: The John Hopkins University Press. Pp. 314-321.

McMahon, David. 2008. *Quantum Field Theory Demystified.* McGraw Hill Professional. Amazon Digital Services, Inc. May 8, 2008. 320 pages.

McTaggart, Lynne. 2009. *The Field: The Quest for the Secret Force of the Universe.* Harper Collins e-books. Amazon Digital Services, Inc. 300 pages.

_____. 2007. *The Intention Experiment: Using Your Thoughts to Change Your Life and the World.* Atria Books. Simon and Schuster

Digital Sales, Inc. Released in Amazon Digital Services, Inc. 336 pages.

Metra J. (ed.) *The Physicist's Conception of Nature*, p. 244.

Mill, John Stuart (1859). *On Liberty*. John W. Parker & Son, West Strand.

Moore, Thomas. 1994. *Care of the Soul*. New York: HarperCollins Publishers.

Morgan, Lloyd. 1923. *Emergent Evolution*. Henru Holt and Co.

Moring, Gary F. 2002. *Theories of the Universe*. N.Y.: Alpha Books.

Musser, George. 2015. *Spooky Action at a Distance: The Phenomenon That Imagines Space and Time—and What it Means for Black Holes, the Big Bang and Theories of Everything*. Amazon Digital Services, Inc. 305 pages.

Nelson, Harry and Robert Jurmain. 1982. *Introduction to Physical Anthropology*. Second edition.

Nelson, Brian. 2014. "Parallel worlds exist and interact with our world, say physicists." November 6, 2014, 7:32 p.m.

Nomad, Ali. 2014. *Cosmic Consciousness: The Man-God Whom We Await*. Amazon Digital Services, Inc. Ali Nomad is the *nom de plume* used by Dr. Alexander J. McIvor-Tyndall, founder of the International New Thought Fellowship.

Nicolson, Iain. 2007. *Dark Side of the Universe: Dark Matter, Dark Energy, and the Fate of the Cosmos*. Baltimore, Maryland: The John Hopkins University Press. 192 pages.

Norquist, Ellwood W. 2014. *Who are We? Science and Spirit Answering With One Voice (Well, Almost)*. Tucson, AZ: Cosmic Connection Publishing. Amazon Digital Services, Inc. 137 pages.

O'Neill, Jennifer. 2013. *Intuition & Psychic Ability: Your Spiritual GPS*. Limitless Publishing LLC. Amazon Digital Services, Inc. 149 pages.

Ornstein, Robert E. 1997. *The Psychology of Consciousness.* New

York: Hartcourt Brace & Company.

Owen, Edgar L. 2013. *Reality: A Sweeping New Vision of the Unity of Existence, Physical Reality, Information, Consciousness, Mind and Time*. USA. Self-published on CreateSpace.com.

Pagels, Heinz R. 2012. *The Cosmic Code: Quantum Physics as the Language of Nature*. Dover Publications. Amazon Digital Services, Inc. 384 pages.

_____. 1990. *The Dreams of Reason: The Computer and the Sciences of Complexity*. Bantam.

Peattie, Lisa R. 1972. "Intervention and Applied Science in Anthropology." Jesse D. Jennings and E. Adamson Hoebel. *Readings in Anthropology*. New York: McGraw-Hill Book Company. Pp. 487-493.

Peleg, Yoav, Reuven Pnini, Elyahu, Zaarur, Eugene Hecht. 2009. *Schaum's Outline of Quantum Mechanics*. McGraw Hill Education. Amazon Digital Services, Inc. 384 pages.

Penrose, Roger. 1989. *The Emperor's Mind: Concerning Computers. Minds, and The Laws of Physics.* New York: Penguin Books.

Perez, Floyd. 2015. *Evolution of Consciousness: A Journey Guided by Universal Law*. Amazon Digital Services, Inc., 369 pages.

Peskin, Michael E. and Daniel V. Schroeder. 1995. *An Introduction to Quantum Field Theory: Frontiers in Physics*. Westview Press. Amazon Digital Services, Inc. 870 pages.

Petrov, Alexey A. and Andrew E. Blechman. 2016. *Effective Field Theories*. Singapore: World Scientific Publishing Co., Pte. Ltd. Amazon Digital Services, Inc.

Piccioni, Robert L. 2013(c). *Quantum Mechanics 3: Wavefunctions, Superposition, & Virtual Particles*. Real Science Publishing. Amazon Digital Services, Inc. 43 pages.

Pinchbeck, Daniel and Ken Jordan. 2012. *Exploring the Edge Realms of Consciousness: Limited Zones, Psychic Science, and the Hidden Dimensions of the Mind*. Evolver Editions. Amazon Digital Services, Inc., 385 pages.

Pribham, Karl. 1976. *Consciousness and the Brain*. New York: Plenum. See also his *Languages of the Brain*. G. Globus et al. (eds.). 1971.

Primack, Joel R. and Nancy Ellen Abrams. 2006. *The View from the Center of the Universe: Discovering Our Extraordinary Place in the Cosmos*. New York: Riverhead Books.

_____. 1995. "In A Beginning..." Quantum Cosmology and Kabbalah. http:// physics.ucsc.edu/cosmo/primack_abrams/ InABeginning Tikkun1995.pdf.

Quill, Elizabeth (ed.). 2016. *Cosmic Frontiers: Scientists Seek Clues to the Universe's Greatest Mysteries*. New York: Diversion Books. Amazon Digital Services, Inc. 284 pages.

Radin, Dean. 2013. *Supernormal: Science, Yoga, and the Evidence for Extraordinary Psychic Abilities*. Random House LLC. Amazon Digital Services, Inc. 401 pages.

_____. 2010. *The Conscious Universe: The Scientific Truth of Psychic Phenomena*. HarperCollins e-books. HarperCollins Publishers. Amazon Digital Services, Inc. 432 pages.

_____. 2006. *Entangled Minds: Extrasensory Experience in a Quantum Reality*. New York: Paraview Pocket Books.

Ranasinghe, Shelton. 2012. *Self-Delusions?* Rocky River: Ohio. Amazon Digital Services, Inc. 48 pages.

Redish, A. David. 2013. *The Mind Within the Brain: How We Make Decisions And How Those Decisions Go Wrong*. Oxford University Press. Amazon Digital Services, Inc. 392 pages.

Rees, Martin. 1997. *Before the Beginning: Our Universe and Others*. Reading, Massachusetts: Addison-Wesley.

Relethford, John H. 1997. *The Human Species: An Introduction to Biological Anthropology*. Mountain View, California: Mayfield Publishing Company. Third edition.

Revonsuo, Antti. 2009. *Consciousness: The Science of Subjectivity*. Psychology Press. Amazon Digital Services, Inc. 353 pages.

Rogers, Buck. 2016. "World's Smartest Physicist Believes

Consciousness Will Remain a Mystery." *Walking Times*. August 23, 2016. Can be accessed at http://www.wakingtimes.com/2016/08/23/worlds-smartest-physicist-believes-consciousness-will-remain-a-mystery.

Rosenblum, Bruce and Fred Kuttner. 2006. *Quantum Enigma: Physics Encounters Consciousness.* N.Y.: Oxford University Press. Amazon Digital Services, Inc.

Rosman, Abraham and Paula G. Rubel. 1998. *The Tapestry of Culture: An Introduction to Cultural Anthropology.* Boston, Massachusetts: McGraw Hill.

Ryder, Lewis H. 1996. *Quantum Field Theory.* Cambridge University Press. Amazon Digital Services, Inc. 507 pages.

Ryle, G. 1949. *The Concept of Mind.* Hutchinson.

Scanlon, Gerarld W. 2015a. *Higgs Field Unveiled God's Field at Creation.* Amazon Digital Services, Inc.

_____. 2015b. *God's Particle and Elements: Core of the Universe – A Theory of Everything.* Amazon Digital Services, Inc.

Scheck, Florian. 2012. *Classical Field Theory on Electrodynamics, Non-Abelian Gauge Theories and Gravitation (Graduate Texts in Physics).* Heidelberg: Springer-Verlag Berlin. Amazon Digital Services, Inc. 436 pages.

Schrodinger, Erwin. 1964. *My View of the World.* Cambridge University Press. Amazon Digital Services, Inc. 118 pages.

Schwartz, Jeffrey M. and Sharon Begley. 2009. *The Mind and the Brain: Neuroplasticity and the Power of Mental Force.* HarperCollins e-books. Harper Collins Publishers. Sold in Kindle Store. 436 pages.

Scott, Love. 2014. *Psychic: Bet You Didn't Have It In You (Psychic, Psychic Development, Psychic Abilities, Clairvoyance, ESP, Channeling, and Mediumship)* Psychic Books. Amazon Digital Services, Inc., 20 pages.

Searle, John R. 1984. *Minds, Brains and Science.* Cambridge, Massachusetts: Harvard University Press.

Seddon, Christopher. 2014. *Humans from the Beginning: From the*

*First Human Apes to the First Cities*. Glanville Subdivisions. Amazon Digital Services, Inc. 567 pages.

Seymour, Percy. 1992. *The Paranormal: Beyond Sensory Science*. Arkana: Penguin Groups. 184 pages.

Shaffer, Lothar. 2013. *Infinite Potential: Quantum Physics Reveal How We Should Live*. Deepak Chopra Publisher. 336 pages. See also his lecture on youtube at http;//www.youtube.com/watch?v=jzafB6NKHis. Accessed July 2014.

Sharp, Michael. 2006-2013.*The Book of Light The Nature of God The Structure of Consciousness, And The Universe Within You*. Alberta, Canada. The Lightning Path Press. Available in .pdf format.

Sheldrake, Rupert. 2012. *The Presence of the Past: Morphic Resonance and the Memory of Nature*. Park Street Press. Amazon Digital Services, Inc. 479 pages.

_____. 2012. *The Presence of the Past: Morphic Resonance and the Memory of Nature*. Park Street Press. Amazon Digital Services, Inc. 479 pages.

_____. 2010. *Morphic Resonance: The Nature of Formative Causation*. Park Street Press. Amazon Digital Services, Inc.

_____. 1995. *The Presence of the Past: Morphic Resonance and the Habits of Nature*. Rochester, Vermont: Park Street Press.

Shubin, Neil. 2013. *The Universe Within: The Deep History of the Human Body*. First Vintage Book Edition. Amazon Digital Services, Inc. 252 pages.

Siegfried, Tom (Managing editor), 2016. "Introduction." In *Cosmic Frontiers: Scientists Seek Clues to the Universe's Greatest Mysteries*. New York: Diversion Books. Amazon Digital Services, Inc. 284 pages.

Sitchin, Zecharia. 1980. *The Stairway to Heaven*. New York: Avon Books.

_____. 1976. *The 12$^{th}$ Planet*. New York: Avon Books.

Speiser, E. A. 1969. *Ancient Near Eastern Texts Relating to the Old Testament*, 3rd edition, edited by James Pritchard (Princeton, 1969).

Splane, Lily. 2004. *Quantum Consciousness: A Philosophy of the Self's Potential Through Quantum Cosmology*. U.S.A.: Anaphase II Publishing.

Sorokin, Pitirim A. 1937-41. *Social and Cultural Dynamics*. Cincinnati: American Book Company.

Srednicki, Mark Allen. 2007. *Quantum Field Theory*. Cambridge University Press. Amazon Digital Services, Inc., 660 pages.

Stager, Curt. 2014. *Your Atomic Self: The Invisible Elements That Connect You To Everything Else in the Universe*. New York: Thomas Dunned Books, St. Martin's Press. Amazon Digital Services, Inc. 321 pages.

Standish, Russell K. 2006. *Theory of Nothing.* Australia: BookSurge. Amazon Digital Services, Inc. 262 pages.

Stapp, Henry P. 2011. *Quantum Reality and Mind.* In Mensky, Michael, Roger Nelson, Horace Crater, Grayson Bruce, George Hathway, Michael Ibison, Kevin Nelson, Michel Cabanac, Jay Kumar, and Deepak Chopra. 2014. *Quantum Physics of the Paranormal: Premonitions, PreCognition, Death, Altered States of Consciousness*. Cosmology Science Publishers. Amazon Digital Services, Inc. 495 pages.

_____. 1995. "Values and the Quantum Conception of Man." Paper presented at UNESCO sponsored symposium "Science and Culture: A Common Pat for the Future," Tokyo, September 1995.

_____. 1971. "Matrix Interpretation of Quantum Theory." *Physical Review*, Vol. D3, pp. 1303-20, March 15, 1971.

Steele, S. and J. Iutcovich, (eds.). 1997. *Directions in Applied Sociology*. Society for Applied Sociology: Arnold, MD. http:/// www.appliedsoc.org/society. Accessed May 16, 2014.

Stringer, Chris. 2012. *Lone Survivors: How We Came to be the Only Humans on Earth.* New York; Times Books. Amazon Digital Services, Inc. 336 pages.

Sundermier, Ali. 2016. "99.9999999% of your body is empty space." *Business Insider*, 23 Sep 2016. Refer to http://www.ferocesmente.com/self-improvement/99-9999999-of-your-body--is-empty-space/.

Susskind, Leonard. 2005. *The Cosmic Landscape: String Theory and the Illusion of Intelligent Design*. New York: Little, Brown & Company.

Swaab D.F. and Jane Hedley-Prole. 2014. *We Are Our Brains: A Neurobiography of the Brain, from the Womb to Alzheimer's*. New York: Penguin Random House Company. Amazon Digital Services, Inc. 448 pages.

Swann, Ingo. 2014. *Penetration: the Question of Extraterrestrial and Human Telepathy*. Cross Road Press. Amazon Digital Services, In. 178 pages.

Tarnas, Richard. 2006. *Cosmos and Psyche: Intimations of a New World View*. New York: Penguin Group. Amazon Digital Services, Inc.

Targ, Russell. 2012. *The Reality of ESP: A Physicist's Proof of Psychic Abilities*. Quest Books. Amazon Digital Services, Inc., 312 pages.

Taylor, Robert Jonathan. 2006. "An Analysis of Celestial Omina in the Light of Mesopotamian Cosmology and Mythos." Thesis Submitted to the Faculty of the Graduate School of Vanderbilt University in partial fulfillment of the requirements for the degree of Master of Arts in Religion. May 2006. Nashville, Tennessee. Refer to http:// etd.library. vanderbilt.edu/ETD-db/available/etd-03312006-193940/unrestricted / final.pdf.

Thich Nhat Hanh. 1995. *Living Buddha, Living Christ*. New York: Riverhead Books.

Tolle, Eckhart. 2004. *The Power of Now: A Guide to Spiritual Enlightenment*. Novato, CA: New World Library.

Trine, Ralph Waldo. 1933. *The Higher Powers of Mind and Spirit*. London: G. Bell and Sons, Ltd.

Tylor, Sir Edward Burnett. 1971. *Primitive Culture: Researches Into the Development of Mythology, Philosophy, Religion, Language, Art, and Custom*. Also published in 2013 by the Amazon Digital Services, Inc. 450 pages.

Tyndall, John. 2012. *Faraday as a Discoverer*. Amazon Digital

Services, Inc. 98 pages.

Tyson, Neil deGrasse. 2007. *Death by Black Hole: And Other Cosmic Quandaries*. W.W. Norton & Company. Amazon Digital Services, Inc., 385 pages.

Uttal, William R. 2011. *Mind and Brain: A Critical Appraisal of Cognitive Science*. Cambridge, Massachusetts London, England: The MIT Press. Amazon Digital Services, Inc. 528 pages.

Van Willigen, John. 1986. *Applied Anthropology*. Massachusetts: Bergin & Garvey Publishers, Inc.

Vilenkin, Alex. 2006. *Many Worlds in One: The Search for Other Universes*. New York: Hill and Wang.

Waldrop, Mitchell M. 1993. *Complexity: The Emerging Science at the Edge of Order and Chaos*. Viking.

Watts, Alan W. 1957. *The Way of Zen*. Vintage. Random House LLC. Also in the Amazon Digital Services, Inc. 256 pages.

West, Peter. 2012. "Introduction to Strings and Branes." New York: Cambridge University Press. Amazon Digital Services, Inc. 723 pages.

Weinberg, Steven. 2015. *Lectures on Quantum Mechanics*. UK: Cambridge University Press. Amazon Digital Services, Inc. 443 pages.

Wheeler, John. 1974. "Beyond the End of Time." In Rees, Ruffini & Wheeler (ed.), *Black Holes, Gravitational Waves and Cosmology*, Gordon and Breach.

Wigner, Eugene. 1967. *Symmetries and Reflections*. Bloomington: Indiana University Press.

Wilcek, Frank. 2015. *A Beautiful Question: Finding Nature's Deep Design*. New York: Penguin Books. Also released in Amazon Digital Services, Inc. 448 pages.

Wilson, Edward O. 1998. *Consilience: The Unity of Knowledge*. Thorndike, Maine: Torndike Press. Also published in 2014 by Vintage Publisher, Random House, LLC. Amazon Digital Services, Inc., 382 pages.

Wolf, Fred Alan. 2010. *Taking the Quantum Leap: The New Physics for Nonscientists*. HarperCollins e-books. Amazon Digital Services, Inc. 306 pages.

_____. 2000. *Mind Into Matter: A New Alchemy of Science and Spirit*. Moment Point Press Inc.. Amazon Digital Services LLC. 188 pages.

_____. 1999. *The Spiritual Universe: One Physicist's Vision of Spirit, Soul, Matter, and Self*. Red Wheel/Weiser. Amazon Digital Services LLC. 325 pages.

_____. 1981. *Taking the Quantum Leap: The New Physics for Nonscientists*. San Francisco: Harper & Row, Publishers.

Wolff, George. 2001. *The BioTech Investor's Bible*. New York: John Wiley & Sons, Inc.

Wolpoff, Milford H. 2000. *Human Evolution*. McGraw Hill.

Wood, Bernard. 2005. *Human Evolution: A Very Short Introduction*. New York: Oxford University Press. Amazon Digital Services, Inc. 131 pages.

Yee, Jeff. 2014. *The Particles of the Universe*. Amazon Digital Services, Inc. 111 pages. 9th edition.

Zee, A. 2010. *Quantum Field Theory in a Nutshell*. New Jersey: Princeton University Press. Amazon Digital Services, Inc. 576 pages.

Zeman, Prof. Adam. 2003. *Consciousness: A User's Guide*. Yale University Press. Amazon Digital Services, Inc. 416 pages.

Zimmer, Heinrich Robert. 1972. *Myths and Symbols in Indian Art and Civilization*. Princeton University Pres. 282 pages.

Zukav G. 2009. *Dancing Wu Li Masters: An Overview of the New Physics*. HarperOne: Later Printing Used Edition 416 pages.

_____. 1990. *The Seat of the Soul*. New York: Simon & Schuster. 255 pages.